UNCHARTED WATERS

UNCHARTED WATERS

The New Economics of Water Scarcity and Variability

Richard Damania, Sébastien Desbureaux,
Marie Hyland, Asif Islam, Scott Moore, Aude-Sophie
Rodella, Jason Russ, and Esha Zaveri

 **WORLD BANK GROUP**

Contents

Boxes

Figures

Maps

Table

Foreword

The 21st century is witnessing the collision of two powerful trends—rising human populations coupled with a changing climate. With population growth, the demand for water is growing exponentially while climate change is making rainfall more erratic and less predictable. Current water management policies are outdated and struggle to address these challenges. If the policy status quo persists, water scarcity will proliferate across new regions of the world and intensify in areas where water is already scarce. This will have far reaching human and economic implications. Water is not only a declared human right, necessary for life and health, but is also a key input into the global economy, powering manufacturing, turning energy turbines, and nourishing crops and livestock. The way in which water is managed in the face of these growing challenges will be key to the economic success of countries where it is scarce and will almost certainly determine whether the world meets its ambitious Sustainable Development Goals.

Managing water is an exceptionally difficult and complex policy challenge. Water is simultaneously a basic right, a natural resource, a fundamental input in all economic activity, and, at times, a source of destruction and devastation. Managing a resource with these multiple and sometimes conflicting attributes presents wicked challenges that lead water experts and policy makers to disagree on how best to manage, protect, and distribute it. It is in this context that this study brings advanced empirical tools and new data to explore the economic consequences of expanding water deficits in the context of a changing climate.

The timing of this study is apt. It builds on the findings and recommendations of the Water Global Practice's previous report *High and Dry*, which was published in May 2016 and explored the water-related impacts of climate change on the global economy. This book also comes one year after the formation of the United Nations' High Level Panel on Water, which called for a fundamental shift in the way the world looks at water.

Uncharted Waters demonstrates the consequences of continuing down the path of the status quo, where water is managed imprudently and urban and rural residents alike are vulnerable to the whims of an increasingly variable climate. But as *Uncharted Waters* demonstrates, these impacts are not far off. Indeed, they are a significant burden in developing countries, with consequences that are often hidden but are nevertheless shockingly large. It is found that, on farms, rainfall variability is responsible for the loss of enough food to feed 81 million people every year. These impacts on farms cascade onto forests, as farmers clear additional land for cultivation to make up for lost yields. In cities, although floods often attract the most media attention, it is drought-like shocks that reduce worker productivity and labor income significantly more than rainfall deluges. Perhaps most tragically, these dry shocks, when experienced during early life, can lead to long-term, unseen impacts on a woman's wealth and education, and even on the health of her offspring.

These impacts demonstrate why, going forward, it is increasingly important that we treat water like the valuable, exhaustible, and degradable resource that it is. This sea change will require a portfolio of policies that acknowledge the economic incentives involved in managing water from its source, to the tap, and back to its source. Constructing new storage and management infrastructure is important but must also be paired with policies that control the demand for water. The utilities that are responsible for water distribution in our cities need to be properly regulated to incentivize better performance and investment in network expansion. And when rainfall shocks turn into economic shocks, safety nets must be put in place to ensure that families can weather the storm.

Meeting the challenges that the future brings will require innovative thinking, unprecedented collaboration, and, perhaps most of all, political will. This study demonstrates the urgency of taking action and provides a roadmap for avoiding the parched path.

Guangzhe Chen
Senior Director
Global Water Practice
World Bank

Acknowledgments

This book was prepared by a team led by Richard Damania and comprising Sébastien Desbureaux, Marie Hyland, Asif Islam, Scott Moore, Aude-Sophie Rodella, Jason Russ, and Esha Zaveri. The book has greatly benefited from the strategic guidance and general direction of the management of the Water Global Practice, including Guangzhe Chen (Senior Director), Jennifer Sara (Director), and Jyoti Shulka (Director).

In addition to research completed by the authors, this work draws on background papers and notes prepared by Dustin Evan Garrick (University of Oxford), Quentin R. Grafton (Australia National University), James Horne (Australia National University), Daniel Camos (World Bank), Antonio Estache (ECARES, Université libre de Bruxelles), and a team at the International Food Policy Research Institute led by Claudia Ringler and comprising Elizabeth Basauri Bryan, Nicostrato D. Perez, Alejandro Nin Pratt, Hua Xie, and Tingju Zhu.

Aleix Serrat Capdevila (World Bank), Laurence Go (University of Pennsylvania), Chuanhao Lin (George Washington University), Anthony Mveyange (World Bank), and Shaffiq Somani (World Bank) provided expert technical assistance, advice, and inputs for which the authors are grateful. The World Bank Water communications and knowledge team—comprising Isabel Hagbrink, Martin Hall, Li Lou, and Pascal Saura—provided invaluable support for turning a manuscript into a finalized book.

The authors received incisive and helpful advice and comments from World Bank colleagues, including Marianne Fay (Chief Economist), Shantayanan Devarajan (Chief Economist), Steven Schonberger (Practice Manager), Uwe Deichmann (Senior Urban Specialist), Somik Lall (Lead Urban Economist), Maria Angelica Sotomayor (Practice Manager), Pilar Maisterra (Practice Manager), Ijsbrand Harko de Jong (Lead Irrigation Specialist), and Greg Browder (Lead Water Resource Management Specialist).

The authors are also indebted to participants of the January 17–18, 2017, Oxford University Workshop for their help in shaping the direction of the research: Felix Pretis, Jacquelyn Pless, Michael Gilmont, Emily Barbour, Johanna Koehler, Jesper Svensson; and to the participants at the panel on Water and the Economy hosted by the Oxford Water Network and the Smith School of Enterprise and the Environment, including Dustin Evan Garrick (Departmental Lecturer, University of Oxford), Jim Hall (Professor, University of Oxford), Simon Dadson (Associate Professor, University of Oxford), and Jean-Paul Penrose (Department for International Development, United Kingdom).

A small team from the formal publishing unit of the World Bank prepared this book: Jewel McFadden (Acquisitions Editor), Rumit Pancholi (Production Editor), and Nora Ridolfi (Print Coordinator), with assistance from Cindy Fisher (Production Editor), and under the supervision of Aziz Gökdemir (Editorial Production Lead) and Patricia Katayama (Acquisitions Lead). Eszter Bodnar designed the front cover and several of the interior images.

This work was made possible by the financial contribution of the Global Water Security and Sanitation Partnership, formerly known as the Water Partnership Program, http://water.worldbank.org/wpp, and the Swiss State Secretariat for Economic Affairs.

Abbreviations

AWD	alternate wetting and drying
CSA	climate-smart agriculture
DHS	Demographic and Health Survey
FAO	Food and Agriculture Organization
GDP	gross domestic product
IFPRI	International Food Policy Research Institute
IMPACT	International Model for Policy Analysis of Agricultural Commodities
IPV	intimate partner violence
ISO	International Organization for Standardization
JADE	Just, Allocative, and Dynamically Efficient
LABLAC	Labor Force Survey Database for Latin America and the Caribbean
NPP	net primary productivity
NTSG	Numerical Terradynamic Simulation Group
PDSI	Palmer Drought Severity Index
PPP	public-private partnership
SEDLAC	Socio-Economic Database for Latin America and the Caribbean
SPEI	Standardized Precipitation Evapotranspiration Index
SPI	Standardized Precipitation Index
SSA	Sub-Saharan Africa
WSS	water supply and sanitation

Executive Summary

When the rains withered and the forests turned into parched savannahs, the earliest humans drifted out of Africa in their quest for water. Farms, settlements, and eventually cities clustered along riverbanks and gave rise to great civilizations. Now, as then, economic activity remains tied to water availability. But this relationship will undergo unprecedented pressures, as the 21st century witnesses the collision of two powerful forces—burgeoning population growth, together with a changing climate. With population growth, water scarcity will proliferate to new areas across the globe. And with climate change, rainfall will become more fickle, with longer and deeper periods of droughts and deluges.

Erratic rains weigh heavily on communities and economies. Floods are so powerful a metaphor of the human experience that nearly every civilization—from classical antiquity, to the Abrahamic religions, to ancient Mesopotamia—tells of a deluge epic that changed the world. Although it is debated whether these myths have a basis in historical events, extreme weather events still reshape societies and permanently mark the lives of those who experience them. Over the past two decades, extreme rainfall events have affected about 300 million people on average every year. With climate change, such extreme episodes of rainfall are expected to increase in frequency. Adapting to changing trends in rainfall, although difficult in its own right, is a gradual and predictable process. Knowing how to address unpredictable rainfall shocks, of uncertain frequency and unknowable magnitude, presents an additional challenge brought by climate change.

Whereas floods are spectacular weather events that cause sensational damage, droughts are misery in slow motion with impacts that are deeper and longer lasting than previously believed. Although overflowing river-banks and storm surges certainly pose major economic threats, this book demonstrates that the impacts of water scarcity and drought may be even greater, causing long-term harm in ways that are poorly understood and inadequately documented. Droughts can have health impacts, hamper firm productivity, accelerate the destruction of forests, and compromise agricultural systems.

This book presents new evidence to advance understanding on how rainfall shocks coupled with water scarcity impact farms, firms, and families. On farms, the largest consumers of water in the world, impacts are

1

channeled from declining yields to changing landscapes. In cities, water extremes, especially when combined with unreliable infrastructure, can stall firm production, sales, and revenue. At the center of this are families, who feel the effects of this uncertainty on their incomes, jobs, and long-term health and welfare.

Parched Farms, Shriveling Yields, and Shrinking Forests

Throughout much of the world, even moderate deviations from normal rainfall levels can cause large changes in crop yields. The driest regions are most sensitive to rainfall variability, although extreme rains can also bring crop losses to regions with more bountiful precipitation and productivity. Such variability is responsible for a considerable net loss of food production every year—enough to feed 81 million people every day, a population the size of Germany's. Many of the affected regions overlap with areas that are already facing large food deficits and are classified as fragile, heightening the urgency of finding and implementing solutions.

Rainfall shocks cascade consequences from declining agricultural yields to shrinking forest cover. Faced with declining agricultural productivity due to rainfall shocks, farmers often seek to recoup these losses by expanding cropland, at the expense of natural habitats. Rainfall variability can account for as much as 60 percent of the increase in the average rate of cropland expansion, and, as a result, is responsible for much of the pressure on forested areas. Climate change may accelerate this pattern, leading to a harmful cycle where rainfall shocks induce deforestation, thereby increasing carbon dioxide emissions, and, in turn, further exacerbating rainfall extremes.

Irrigation systems usually insulate agriculture from the adverse effects of rainfall variability, but these systems may also paradoxically amplify the impacts of shocks. The availability of irrigation typically provides both a buffer against rainfall variability and a significant boost to crop yields in normal years. However, in many dry regions of the world these systems fail to protect farmers from the impacts of droughts. Free irrigation water creates the illusion of abundance, which buoys the cultivation of water-intensive crops such as rice and sugarcane that are ultimately unsuited to these regions. The ironclad laws of demand and supply then dictate that when water is provided too cheaply, it is also consumed recklessly. As a result, crop productivity suffers disproportionately in times of dry shocks due to extraordinary water needs that cannot be met. This book demonstrates that this *paradox of supply* is a widespread problem in areas where water is scarce and its demand is uncontrolled.

When Rainfall Becomes Destiny

Although a rainfall shock may be fleeting, its consequences can shape the destiny of those who experience it in infancy. Deprivations, such as a lack of food, endured in early life impede the physical and mental development of a child with significant and often irreversible consequences.

In rural Africa, women born during severe droughts bear the marks throughout their lives, growing up physically shorter, receiving less education, and ultimately, becoming less wealthy. They may also be less empowered to make household financial decisions and more accepting of domestic violence. Droughts tend to be viewed as short-term events that end as soon as the rains start falling again, but their effects can haunt individuals throughout their lives, causing impacts that go undetected.

Perhaps most troubling, the legacy of rainfall shocks can ripple through generations, harming not just the women who experienced them, but also their children. Rainfall shocks experienced by a mother in her own infancy can significantly impact the health of her children, who are more likely to suffer from malnutrition. These findings add to the urgency of addressing the effects of adversity in infancy through steps such as drought insurance or social safety nets.

Vulnerability in the City

In cities, the economic impacts of a dry shock are often greater than those of a wet shock. While urban infrastructure is generally able to buffer residents against the effects of moderate rainfall shocks, cities are still at the mercy of large rainfall shocks. Further, while the immediate devastation caused by floods attracts much attention, droughts in cities may have the longer-lasting, more severe impact on firms and their employees. In Latin America, losses in income caused by a dry shock are four times greater than that of a wet shock. Droughts have poorly understood consequences within cities, causing higher incidences of diarrheal diseases, health impacts on young children, and an increased frequency of power outages.

The performance of firms in cities is also affected by the availability of water. While the private sector's reliance on transport and energy infrastructure is well established, little is known about the significance of water to firms. Findings in this book show that when urban water services are disrupted, whether by climate, inadequate infrastructure, or both, firms suffer significant reductions in their sales and employment. Particularly vulnerable are small and informal firms, a major source of employment in developing countries. The impacts of water supply and sanitation services in cities therefore extend beyond the widely documented effects on human health.

Avoiding the Parched Path

Pursuing business as usual will lead many countries down a "parched path" where droughts shape destinies. Avoiding this misery in slow motion will call for fundamental changes to how water is managed. It will require using different policy instruments to address the multiple economic attributes of water, through its cycle of use (figure ES.1).

At its source, in rivers, forests, and aquifers, water is a public good subject to all the mismanagement and overexploitation problems of a common-property resource. As water moves into pipes to quench the thirst of cities, or into irrigation canals to grow food, it becomes simultaneously a private good and a merit good—one to which people have a right as a necessity for life and health. In cities, this dual challenge is compounded by the fact that the costs of building multiple water

FIGURE ES.1 **The Water Policy Cycle**

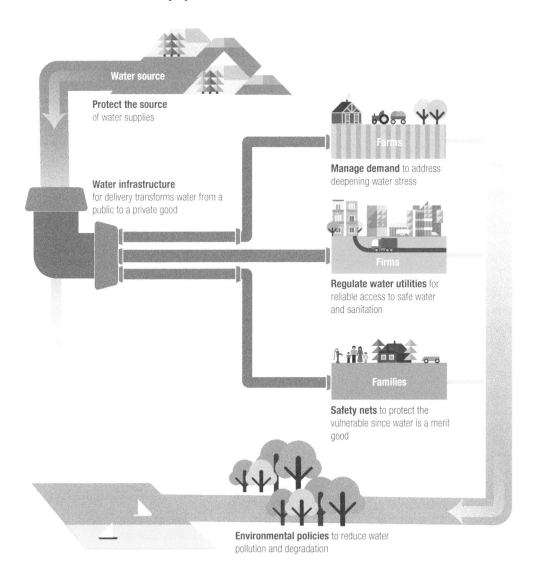

Water source

Protect the source of water supplies

Farms

Manage demand to address deepening water stress

Water infrastructure for delivery transforms water from a public to a private good

Firms

Regulate water utilities for reliable access to safe water and sanitation

Families

Safety nets to protect the vulnerable since water is a merit good

Environmental policies to reduce water pollution and degradation

systems is prohibitive and impractical. Water must therefore be supplied to consumers through a single network, which must have a single owner—a monopolist—that needs to be regulated to ensure adequate access to water at a price that people can afford. Finally, the water passes through sewers and reenters the ecosystem where, if untreated, it can pose major health and environmental risks. These multiple, and at times competing, attributes often cause policy makers, economists, environmentalists, and water experts alike to disagree on how best to regulate, distribute, and use water. But neglecting these linkages can result in policy decisions that are at best less effective than they could be, and at worst downright harmful.

At its source, supply-side measures are needed to deliver water to users. These may encompass investments in infrastructure like reservoirs, irrigation systems, and wastewater reuse technologies. "Natural capital" solutions, which draw on such features of nature as the water-retaining abilities of forests, offer relatively inexpensive means of addressing some water scarcity issues as well. Investments in technologies that improve the efficiency of water used and consumed may be helpful too and offer the tantalizing prospect of creating "new" water, without depriving any existing users. Adoption of these solutions has been slow due to misaligned incentives. A large proportion of the benefits of efficiency improvements are public, while technology adoption costs are private. This implies that sharper incentives are required for technology uptake that might include a change in the subsidy regime, improved access to credit, or public investments in infrastructure. The challenge of managing water in cities is fundamentally different, demanding better performance from water utilities through appropriate forms of regulation and incentives to assure a balance between quality services for consumers, and a rate of return that assures cost recovery and further investment in the sector.

Supply-side approaches, though necessary, are seldom sufficient to build adequate resilience to fickle rainfall patterns. Without proper economic signals, increasing the supply of water often also increases demand. The result is a vicious cycle where water supplies are expanded only to see that water consumed inefficiently; eventually returning the region to worsening levels of water stress. This *paradox of supply* forcefully illustrates the need for combining investments that expand water supplies with policies that manage demand and allocate water efficiently. Such policies include water pricing, water trading exchanges, and quotas on overall water use to ensure enough is left over for the environment. Water trading schemes are a promising approach that allows for the sale of water to higher-valued uses. The result is a win-win because a transfer occurs only if buyer and seller both benefit from the transaction. Efficiency of use rises and conservation improves. The institutional architecture required for a well-functioning water trading system is complex. But even if the obstacles seem significant, this is an instrument whose time has come for consideration, if not immediate implementation in all contexts.

Improved management in the water sector, while necessary for building efficiency and resilience, may not protect the poor from erratic rains nor assure that water is used sustainably. Safety net programs and insurance schemes are needed to protect the most vulnerable populations from the torments of droughts and floods. In rural areas, these safety nets could take the form of crop insurance schemes, while in cities, careful utility regulation is needed to ensure affordable access to clean water. Adequate safeguards, such as quotas and water quality standards, are required to ensure more sustainable water use, to protect water sources, and to prevent over-use and abuse of these public goods. This mix of policy tools is needed to protect those most vulnerable to water shocks, and ensure that rainfall does not become destiny, perpetuating poverty.

The future will be thirsty and uncertain. Already more than 60 percent of humanity live in areas of water stress where available supplies cannot sustainably meet demand. If water is not managed more prudently—from source, to tap, and back to source—the crises observed today will become the catastrophes of tomorrow.

— UNCHARTED WATERS —

With population growth the **demand for water is accelerating** and with climate change **rainfall has become more erratic**

Rainfall shocks affect about **25%** of humanity each year

Impacts ripple across farms, firms, and families

FARMS	FIRMS	FAMILIES
Dry shocks reduce yields and cause annual losses that could feed 81 million people, the population of Germany. Dry shocks push farmers to expand agriculture into forests, worsening climate change and threatening water supplies	For firms and cities, the cost of dry shocks are four times greater than wet shocks. Without sufficient water, economies slow down with impacts on health, labor incomes, and firm sales	A dry shock in infancy can become destiny, with lasting effects on health and wealth, trapping subsequent generations in poverty and malnutrition

Dry shocks are **misery in slow motion**

1
Water in the Balance

"For only what is rare is valuable; and water, which, as Pindar says, is the 'best of all things,' is also the cheapest."

—Plato, *Euthydemus*

Introduction

This book is about the economics of water. Few resources on this planet are more important than water for human health and economic activity, and few are more complex. Understanding the economics of water has long challenged scholars, philosophers, and policy makers.

In 1776, Adam Smith observed that water, although essential for life and production, is cheap, while diamonds, a mere adornment, are exorbitantly priced. Termed the *Water-Diamond Paradox*, it seemed irrational that an economy could place a higher value on a jewel that serves no productive purpose than on a resource that is a necessity. The origins of the paradox are even more ancient and appear in one of Plato's dialogues (*Euthydemus*).

The textbook response to the Water-Diamond Paradox is that prices mirror scarcity: Diamonds are rare and expensive to extract, so buyers must pay a high price for small quantities of glitter. Water, though indispensable, is plentiful, so it commands a much lower price. The paradox has emerged as the textbook staple for teaching the importance of scarcity in economics.

But experience suggests that this alluring textbook response is at best only partially accurate, if not somewhat misleading. Adam Smith, the 18th-century professor of moral philosophy, lived in wet, sparsely populated Scotland and never considered the possibility of water being a scarce and constraining resource. Today, swelling populations, rising demand for water and its related products, and a changing climate have wiped out

---

The technical working papers showing full results are available in volume 2 at www .worldbank.org/UnchartedWaters.

water surpluses even in some of the wetter regions of the globe, and have parched some of its already-dry regions.

Nevertheless, examples abound of countries where water is scarce and yet is used more intensively and wastefully than in countries where it is abundant (figure 1.1). In many of these countries, water is sold at artificially cheap prices, often with the intention of ensuring affordability. But this also means that consumers have little incentive to conserve it and investors have little incentive to lay pipes and deliver it to those who need it. In addition, the pollution plumes of cities, agriculture, and industry have emerged as a serious health hazard and further threaten the sustainability of water supplies. Thus, neither markets nor government policies appear to be allocating water in ways that promote efficiency and assure sustainability for future generations.

Today, the increasing variability and uncertainty of rainfall brought about by climate change are magnifying these challenges. *Rainfall shocks*, where rainfall levels are well above or below normal levels, are becoming much more frequent, and coping with them may present one of the most difficult challenges of climate change. On the other hand, adapting to changing *averages* of climate, though difficult in its own right, is a gradual and more forecastable process. Conversely, increased variability brings uncertainty and unpredictability, which render typical adaptation strategies ineffective.

FIGURE 1.1 **Comparison of Water-Intensive and Water-Scarce Economies**

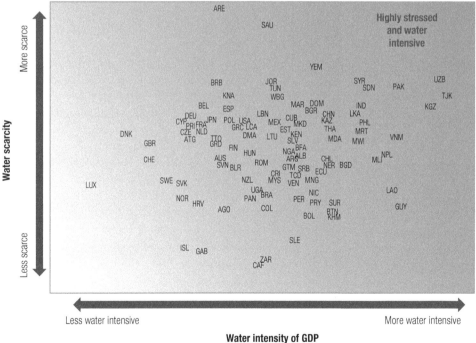

Sources: Water data: AQUASTAT database of the Food and Agriculture Organization (FAO); Output data: World Bank World Development Indicators.
Note: Figure 1.1 compares water intensity of GDP with water scarcity. Water intensity of GDP is measured as the ratio of total economic output to total water withdrawals, and water scarcity is measured as the ratio of total water withdrawals to renewable freshwater resources. Country abbreviations are ISO codes.

A Complex Natural Resource That Presents
Wicked Challenges

Solutions to managing water variability and scarcity remain elusive, despite vast investments in infrastructure and generous funding for research. Multifaceted in its uses, water is allocated in ways that are unequal, both by nature and by human design, and its distribution is changing and becoming less predictable because of climate change. It embodies five often conflicting attributes that render its management especially difficult: It is simultaneously a merit good, a public good, an exhaustible resource, a private good (sometimes), and, during its conveyance, a natural monopoly (box 1.1).

BOX 1.1. **The Economic Attributes of Water**

- Water is deemed to be a merit good—that is, something to which people have a right, regardless of ability to pay—because it is essential for life. It is enshrined as a human right in Resolution 64/292 of the UN General Assembly,[a] which calls upon governments to ensure adequate and affordable quantities of safe water for domestic use.

- At its source, watersheds and rivers, water embodies many of the characteristics of common-property public goods that are vulnerable to the tragedy-of-the-commons problem, which may lead to overexploitation of a shared resource.

- As an exhaustible resource (that is, one that may be renewable, but is depletable), problems of pollution, overexploitation, overuse, and abuse could diminish or destroy the productivity of the resource.

- When water reaches a faucet, it is a private good, since the consumption of water by one user precludes its use by others. At the same time, however, some users merely "rent" the resource—for example, a hydropower plant which uses water to turn a turbine before returning it to the system—while other users "buy" the resource and consume it, such that it is lost from the system, as with crops that absorb and retain the water. The ability to tell the difference between the proportions that are "bought" (not returned to the system) and "rented" (returned to the basin in an undiminished quality) is important, yet often difficult.

- Last, establishing networks of pipes and sewage systems requires large capital investments. As a result, the most cost-effective way to supply water to consumers is through a single pipe, which in turn must have a single owner—a monopolist. This brings the risk that monopolists could exploit their market power by inflating costs and prices.

a. http://www.un.org/es/comun/docs/?symbol=A/RES/64/292&lang=E.

Collectively, these factors render the management of water a challenging endeavor. A policy or price that makes access to water more affordable may condone profligacy and compromise sustainability. At the same time, the economic imperative to price water to cover costs may expose consumers to the exploitive powers of a monopolist. Left unregulated, there would be incentives for users to overexploit and pollute water sources.

If water systems are to meet multiple objectives simultaneously, separate policy instruments will be needed to address each conflicting goal. Where, for instance, water is priced to promote economic efficiency, concerns about affordability may call for compensating those in need with subsidies, while environmental imperatives would necessitate quotas on water withdrawals combined with water-quality standards. In sum, the competing objectives of water management frequently require multiple and complementary policies.

A Global Backdrop of Expanding Water Deficits and Declining Water Quality

Projections suggest that by 2050, global demand for water will increase by 30–50 percent, driven by population growth, rising consumption, urbanization, and energy needs. At the same time, water supplies are limited and under stress from negligent management, growing pollution, degraded watersheds, and climate change.

As many as 4 billion people already live in regions that experience severe water stress for at least part of the year (Mekonnen and Hoekstra 2016). With populations rising, these stresses will mount. Many regions that are likely to see the highest rates of population growth by 2050 are already the most water-stressed and impoverished and will likely endure even greater water deficits (map 1.1).[1]

The expansion of global agricultural production is one of the great economic success stories of our time. The world produces more than enough food to meet its needs and wastes much of it, though equitable distribution remains a challenge. However, this tremendous accomplishment masks unintended impacts on water supplies. First, agricultural growth has often been achieved through increased irrigation and destruction of natural habitats, which can accentuate water scarcities and deplete the very natural resource base on which the sector depends. Second, along with the productivity increase has come a rise in fertilizer use. Estimates suggest that about 50 percent of the yield increases of the past century are due to greater use of fertilizers and irrigation (Erisman et al. 2008). An unintended consequence has been vast pollution plumes in waterways. Nutrient runoff from fertilized cropland releases large amounts of nitrogen and phosphorous into the environment. While these elements are essential for sustaining life, excessive concentrations of nitrogen and phosphorus in water result in eutrophication, a situation in which certain plants and algae "bloom" en masse, outcompeting other species and depriving aquatic life of oxygen. In addition to devastating local ecosystems, this exacts an economic and social cost. The consumption of eutrophic water

MAP 1.1 **Per Capita Water Availability and Future Population Growth, 2050**

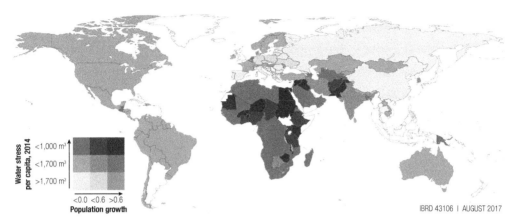

Sources: Freshwater availability data: FAO AQUASTAT database. Population growth estimates: United Nations Population Division, World Population Prospects, 2015 revision (moderate scenario), for the year 2050.
Note: Map 1.1 shows the intersection of water stress, measured as per capita water availability, and population growth. Data on water availability are missing for countries in white.

MAP 1.2 **Water Scarcity and Water Pollution, 2014–15**

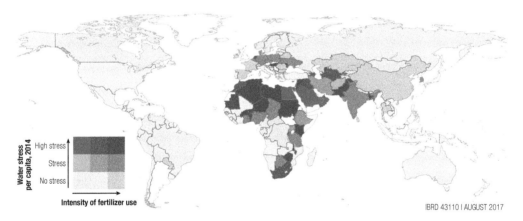

Sources: Freshwater availability data are from FAO's AQUASTAT database. Water pollution data were calculated by the International Food Policy Research Institute (IFPRI) using the International Model for Policy Analysis of Agricultural Commodities and Trade (IMPACT) model and national statistics on livestock and crop areas; these inputs are fed into a hydrological model to estimate nitrogen and phosphorus loading densities.
Note: Map 1.2 shows the intersection of renewable freshwater availability per capita and water pollution due to fertilizer use. Data on water stress are missing for countries in white.

creates health problems, ranging from fatigue and headaches to diarrhea and organ damage (World Health Organization 2002). Eutrophication also takes a heavy toll on the seafood industry, killing fish in huge quantities and reducing yields. There is a growing scientific consensus that the hypothesized safe boundary for the global nitrogen cycle has been crossed (Rockström et al. 2009). As with population growth, some of the most polluted waters due to fertilizer runoff are found in water-stressed regions (map 1.2). This suggests that the business-as-usual path to agricultural growth may not be sustainable for feeding a global population that is projected to exceed more than 9 billion people by 2050.

The Increasing Variability of Rainfall

Water scarcity, stress, and climate change are typically portrayed through a lens of averages and trends. But this is seldom an adequate representation of water availability throughout much of the world, where deviations from trends are widespread and are growing more frequent, as witnessed by the increased frequency of floods and droughts.

Adapting to rainfall variability is often much more challenging than accommodating long-term trends because of the unpredictable duration of a deviation, its uncertain magnitude, and its unknown frequency (Adams et al. 2013). With climate change, deviations from trends are projected to become more pronounced and more frequent. Inter-annual variability in particular is expected to pose a large threat in some of the world's driest regions, including the southwestern United States, Australia, the Middle East, North Africa, and Central Asia (Hall et al. 2014). There is considerable uncertainty around changes in the distribution of rainfall across and between dry and wet regions. Some studies have found that in certain geographies, areas with water surpluses are becoming drier, and conversely some drier areas are becoming wetter (Ashfaq et al. 2009; Chaturvedi et al. 2012; Donat et al. 2016; Ghosh et al. 2016; Greve et al. 2014; Hu et al. 2000; Krishnan et al. 2016). Significantly, however, most models do suggest that rainfall variability will increase (Hall et al. 2014).

Addressing the worsening problem of rainfall variability is not a distant challenge for the future. Much of the world already suffers from inter-annual variation in rainfall. Over the last three decades, 1.8 billion people, or approximately 25 percent of humanity, have endured abnormal rainfall episodes each year, whether it was a particularly wet year or an unusually dry one. Over this period, 300 million people every year have had to cope with destructive rainfall events that are supposed to be extremely rare, the sort of event that might be expected twice in a century in a given location.[2] Unfortunately, variability has disproportionately impacted developing nations, with upward of 85 percent of affected people living in low- or middle-income countries. This book tracks the significant economic and social disruption that these episodes cause, particularly in places that lack the necessary infrastructure to buffer the shocks.

A Snapshot of the Approach

This book uses a variety of approaches to advance the state of knowledge on how water-related challenges impact farms, firms, and families. The issues relating to the economic impacts of water are wide-ranging, their ramifications are endless, and the knowledge gaps are enormous. Addressing all of these challenges is beyond the scope of this book. Instead, the primary, though not exclusive, focus of this book is on the effects of rainfall variability—because of their presumed impacts and the difficulties of adapting to uncertain and unpredictable changes. Farms, which rely on rainfall and water from irrigation, are already facing increasing uncertainty of rainfall, and often find it hard to adapt. Firms—many of which require a large, reliable, and clean water supply—are hampered by deteriorating and inadequate infrastructure in cities that are becoming ever thirstier as they grow. And at the center of this, families, both in cities and in rural areas, rely on increasingly stressed, polluted, and variable water resources both for survival and the livelihoods that come from farms and firms.

For analytical purposes, variations in rainfall are measured and characterized in this book in terms of a statistical (that is, a "standard") deviation of rainfall from its long-term mean. Definitions of floods and droughts vary and are often context-specific, so the deviations analyzed in this book may not always be classified as either a drought or a flood. As an example, an unusually intense downpour may not cause flooding in a city that is located on a hill or is endowed with a modern and efficient drainage system. But the same downpour may have catastrophic impacts in another area that has inadequate drainage and more challenging topography. More information on these issues is provided in appendix A.

The chapters that follow in volume 1 of this book describe the key findings of the analyses, with technical details relegated to volume 2, which is available online. Chapter 2 examines an old and recurring concern—the impacts of rainfall variability on global food production. In order to do so, the impacts of rainfall shocks on food production along two margins are examined: the intensive margin, or how yields are impacted, and the extensive margin, or how cropping patterns and land-use dynamics are affected. The chapter also examines how large-scale irrigation infrastructure influences these two margins and mitigates the effects of rainfall variability.

Chapter 3 explores the human impacts of rainfall shocks. It finds that in rural areas, living through a long-term drought in childhood can leave scars that follow individuals throughout their lives. Stresses caused by a *single* severe dry shock can impair cognitive and physical development and have lasting harmful effects that are even transmitted to future generations.

Chapter 4 investigates rainfall shocks and cities, and asks whether economic growth and urban labor markets react to these events. Cities are shown to be vulnerable to extreme shocks—both wet and dry—with particularly large consequences for droughts, even in middle-income countries. While the impacts of electricity and transportation on firm performance are well understood, there is little information on whether firm performance is also affected by water infrastructure. This chapter further explores this issue and finds that access to a reliable source of piped water is critical for business performance.

Finally, chapter 5 builds on the results of this book to offer practical guidelines for policy makers to address the water challenges of the 21st century.

Notes

1. Note that the countries showing on this map with low water availability per capita do not line up perfectly with the water-stressed countries shown in figure 2.1. There are many different ways to measure water stress and water availability. Water per capita will take into account only water endowments and demographics, while the water stress measure used in figure 2.1 also considers how intensively water is used.

2. This statement assumes a normal distribution of rainfall.

References

Adams, S., F. Baarsch, A. Bondeau, D. Coumou, R. Donner, K. Frieler, B. Hare, A. Menon, M. Perette, F. Piontek, K. Rehfeld, A. Robinson, M. Rocha, J. Rogelj, J. Runge, M. Schaeffer, J. Schewe, C.-F. Schleussner, S. Schwan, O. Serdeczny, A. Svirejeva-Hopkins, M. Vieweg, and L. Warszawski. 2013. *Turn Down the Heat: Climate Extremes, Regional Impacts, and the Case for Resilience—Full Report*. Washington, DC: World Bank.

Ashfaq, M., Y. Shi, W. Tung, R. J. Trapp, X. Gao, J. S. Pal, and N. S. Diffenbaugh. 2009. "Suppression of South Asian Summer Monsoon Precipitation in the 21st Century." *Geophysical Research Letters* 36 (1): L01704.

Chaturvedi, R. K., J. Joshi, M. Jayaraman, G. Bala, and N. Ravindranath. 2012. "Multi-Model Climate Change Projections for India under Representative Concentration Pathways." *Current Science* 103 (7): 791–802.

Donat, M., A. Lowry, L. Alexander, P. O'Gorman, and N. Maher. 2016. "More Extreme Precipitation in the World's Dry and Wet Regions." *Nature Climate Change* 6: 508–13.

Erisman, J., M. Sutton, J. Galloway, Z. Klimont, and W. Winiwarter. 2008. "How a Century of Ammonia Synthesis Changed the World." *Nature Geoscience* 1 (10): 636–39.

Food and Agriculture Organization (FAO). AQUASTAT database. FAO, Rome. http://www.fao.org/nr/water/aquastat/countries_regions/index.stm.

Ghosh, S., H. Vittal, T. Sharma, S. Karmakar, K. Kasiviswanathan, Y. Dhanesh, K. Sudheer, and S. Gunthe. 2016. "Indian Summer Monsoon Rainfall: Implications of Contrasting Trends in the Spatial Variability of Means and Extremes." *PLoS One* 11 (7): e0158670.

Greve, P., B. Orlowsky, B. Mueller, J. Sheffield, M. Reichstein, and S. I. Seneviratne. 2014. "Global Assessment of Trends in Wetting and Drying over Land." *Nature Geoscience* 7 (10): 716–21.

Hall, J. W., D. Grey, D. Garrick, F. Fung, C. Brown, S. J. Dadson, and C. W. Sadoff. 2014. "Coping with the Curse of Freshwater Variability." *Science* 346 (6208): 429–30.

Hu, Z. Z., Latif, M., Roeckner, E., and Bengtsson, L. 2000. "Intensified Asian Summer Monsoon and Its Variability in a Coupled Model Forced by Increasing Greenhouse Gas Concentrations." *Geophysical Research Letters* 27 (17): 2681–84.

Krishnan, R., T. P. Sabin, R. Vellore, M. Mujumdar, J. Sanjay, B. N. Goswami, F. Hourdin, J.-L. Dufresne, and P. Terray. 2016. "Deciphering the Desiccation Trend of the South Asian Monsoon Hydroclimate in a Warming World." *Climate Dynamics* 47 (3–4): 1007–27.

Mekonnen, M., and A. Hoekstra. 2016. "Four Billion People Facing Severe Water Scarcity." *Science Advances* 2 (2): e1500323.

Rockström, J., W. Steffen, K. Noone, Å. Persson, F. S. Chapin III, E. F. Lambin, T. M. Lenton, M. Scheffer, C. Folke, H. J. Schellnhuber, B. Nykvist, C. A. de Wit, T. Hughes, S. van der Leeuw, H. Rodhe, S. Sörlin, P. K. Snyder, R. Costanza, U. Svedin, M. Falkenmark,

L. Karlberg, R. W. Corell, V. J. Fabry, J. Hansen, B. Walker, D. Liverman, K. Richardson, P. Crutzen, and J. A. Foley. 2009. "A Safe Operating Space for Humanity." *Nature* 461 (7263): 472–75.

World Bank. World Development Indicators database. World Bank, Washington, DC. http://data.worldbank.org/data-catalog/world-development-indicators.

World Health Organization. 2002. "Eutrophication and Health." Luxembourg: Office for Official Publications of the European Communities.

World Population Division, United Nations. 2015. World Population Prospects, 2015 revision (moderate scenario), for the year 2050. https://esa.un.org/unpd/wpp /Download/Standard/Population/.

2
Drenched Fields and Parched Farms

As between cultivable land and land with mines, cultivable land is preferable. For mines only fill the treasury, while grains fill both the treasury with taxes, and the storehouses with produce.

—Kautilya, 4th-century BCE scholar, in the treatise *Arthashatra*

Key Chapter Findings

- Rainfall shocks have a significant impact on crop productivity, with even moderately dry shocks reducing productivity and wet shocks increasing productivity.
- Since 2001, rainfall shocks have caused a loss of food calories sufficient to feed about 81 million people each year.
- In low- and middle-income countries, dry shocks can lead to accelerated cropland expansion at the expense of forests, as farmers try to recoup productivity losses by increasing the amount of land that they cultivate.
- Irrigation infrastructure has the potential to buffer crops against these losses, and eliminate the need for farmers to expand their cropland.
- But a paradox of supply may prevail: In arid areas, free irrigation water can induce maladaptation, whereby farmers grow water-intensive crops that accentuate their vulnerability to drought. Crop productivity then suffers disproportionately in times of dry shocks as a result of the unmet extraordinary water needs, and impacts are worsened.

In perhaps no other sector is water access, reliability, and predictability more important for economic success than in agriculture. This has implications far exceeding mere profits and losses, with food security, proper nutrition, rural livelihoods, and environmental sustainability all depending on water. This chapter examines how rainfall shocks impact crop production and seeks to answer three compelling questions: To what extent are global agricultural productivity and global food security sensitive to rainfall variability? What is the relationship, if any, between rainfall variability and land-use changes and, by implication, deforestation? And finally, how effective has irrigation infrastructure been as a buffer against rainfall variability? More broadly, this chapter also examines whether infrastructure investment will be a viable strategy for adapting to climate change and meeting the world's increasing food requirements during the 21st century.

The Challenge of Rainfall Variability

For thousands of years, humans have struggled to adapt to the unpredictable nature of rainfall variations. Indeed, written accounts dealing with rainfall fluctuations date back to the ancient treatise *Arthashatra*, written in 4th-century BCE by the Indian scholar Kautilya, who discussed ways to predict and adapt to erratic monsoon rains.[1] This issue continues to occupy center stage in policy discussions on food security. Looking into the future, the problem could worsen with two 21st-century transitions—growing populations that propel an increase in the demand for food and water, coupled with a changing climate that renders rainfall more erratic and less predictable. A deeper understanding of the consequences of these trends and the effectiveness of remedies that buffer economies from rainfall variability will be helpful in finding more effective responses to these problems.

These issues are of far-reaching policy significance for at least three reasons. First, if adverse rainfall shocks disrupt agriculture in a significant way, then the consequences could cascade to other parts of the economy with wider implications. Second, and of greater concern, rainfall variability is greatest in many of the poorest countries of the world, with rapidly expanding populations and elevated levels of water stress, which compromises their ability to meet escalating domestic food demand. Third, there could be environmental implications that impact the sustainability of food production. The current common practice of increasing food supply by expanding the land area under cultivation—termed extensification—is rapidly becoming unsustainable as land grows scarcer and forest, rangeland, and wetland margins are exhausted.

Much to Be Learned, Despite Decades of Research

The existing literature on the effects of rainfall shocks on agriculture is surprisingly ambiguous, and as a result offers limited policy advice on how best to deal with these issues. Crop-specific agricultural studies convincingly demonstrate the effects of rainfall on agricultural yields. Such research spans the globe, ranging from rice yields in India (Auffhammer et al. 2006),

Indonesia (Naylor et al. 2001), and Thailand (Sawano et al. 2008), to maize (corn) in the United States (Lobell et al. 2013) and soybeans in Argentina (Verón et al. 2015). Studies have established too that what matters is not just the total level of rainfall, but also its variability and distribution throughout the year, which can have a much larger impact (Fishman 2016).

However, the literature on the economic effects of climate change is much more equivocal.[2] There is a broad consensus in the literature that higher temperatures induce strong and lasting negative impacts on economic performance. However, most studies find no discernible evidence that rainfall has any consistent impact on economic activity, including aggregate agricultural value added (Burke, Hsiang, and Miguel 2015; Dell, Jones, and Olken 2012).

There could be many factors that might explain this ambiguity. But this chapter demonstrates that the level of spatial disaggregation is key in reconciling these contradictory results.[3] When the effects of rainfall are measured at a sufficiently disaggregated spatial scale, significant impacts are found on agriculture that vary considerably across agroclimatic regions.

Agricultural Productivity Is Highly Sensitive to Rainfall Shocks

The analysis in this chapter is based on disaggregated data sets of crop productivity, cropland, and other variables that cover 175 countries. The focus is on the effects of rainfall episodes that deviate at least modestly (that is, by more than 1 standard deviation, which on average is a 28 percent deviation), from the long-term average level of rainfall in that region. In global terms, this equates to a deviation of 133 millimeters (mm) in rainfall relative to a 700 mm mean (which is the global average). These "shocks" represent largely unpredictable and hence exogenous variations in rainfall since they are found to be random (in the sense that the series are stationary and white noise).[4] Such episodes tend to occur on average between one and two times per decade. Box 2.1 describes the methodology and some of the data employed in more detail.

Relatively moderate deviations away from long-run averages cause large changes in crop production as measured using net primary productivity, which is a common metric of agricultural output that is available at a global scale (box 2.1). There is a predictable difference between wet shocks and dry shocks. On average, a dry shock (defined as rainfall that is at least 1 standard deviation below normal levels) can reduce agricultural productivity by approximately 10 percent. On the other hand, wet shocks (rainfall levels that are at least 1 standard deviation higher than normal) tend to increase agricultural productivity by approximately 7.4 percent.

These global averages mask considerable regional variation. Figure 2.1 shows that the driest regions (on the left) are the most sensitive to rainfall variability. In these areas, a wet shock can increase crop yields by nearly 14 percent, while a dry shock can decrease them by 17 percent. In wetter

BOX 2.1. A Look under the Hood at the Data and Methodology Employed

The impact of rainfall variability on agricultural productivity and cropland expansion is assessed using a variety of statistical techniques, which are outlined in volume 2 of this book. Global land area is split into grid cells measuring 0.5 degree on each side. Net primary productivity (NPP) is used as the preferred measure of agricultural performance for a number of reasons (Running et al. 2004). NPP, which is linearly related to the amount of solar energy that plants absorb over a growing season, measures the amount of carbon stored for food consumption per square meter. It therefore serves as the principal energy source for crops and, in turn, for human populations that depend upon them (Abdi et al. 2014).[a] NPP has the virtue of being convertible into kilocalories (Blanc and Strobl 2013). As a proxy for nutritional value, and can be measured from satellite imagery, thereby reducing possible measurement errors and providing observations for the entire planet for the period of study, 2000–13. It also provides a common unit of measurement across different crops that is convertible directly into a measure of food availability (Lobell et al. 2002; Lu and Zhuang 2010; Tum and Gunther 2011).

Combining NPP with a data set on cropland, total annual NPP from agriculture was aggregated into approximately 9,300 grid cells in 175 countries over the time period. Cropland data were sourced from Earthstat global cropland data, which were developed by the Land Use and Global Environment Research Group at McGill University and are available from 1980 to 2005 at five-year intervals.

Rainfall variability is measured in terms of local deviations from the long-run mean (annual average rainfall from 1900 to 2014). Specifically, a grid cell is considered to have a *dry rainfall shock* if rainfall in a given year is lower than the long-run annual mean for the grid cell by at least 1 standard deviation. Similarly, a grid cell is considered to have a *wet rainfall shock* if rainfall in a given year is higher than the long-run annual mean for the grid cell by at least 1 standard deviation. See appendix A for more information on the rainfall data.

The relationship between the incidence of these shocks and the change in NPP is estimated using a variety of regression techniques. Control variables, including temperature and various fixed effects and time trends, are included to ensure that the impact of rainfall shocks is estimated precisely and not biased by factors that may be correlated with both agricultural production and rainfall shocks.

To estimate the impact of rainfall shocks on cropland expansion, this setup is modified to examine medium-term outcomes. The expansion of cropland, particularly onto virgin fields or into forested areas, can require a large, upfront fixed cost. For this reason, one might expect to see little immediate response to a single variable rainfall event, but that repeated shocks over the medium term could induce farmers to expand their fields as an adaptation strategy.

The rainfall shock variables are therefore modified to measure the *number* of years in the previous 10 years for which rainfall was at least 1 standard deviation above or below the mean. In subsequent chapters, a similar methodology is used to estimate the impact of rainfall shocks on household wealth, socioeconomic factors, and urban labor outcomes.

a. Note that NPP only measures crop productivity and does not account for livestock. As there is no global panel data set on livestock production, this is omitted from the analysis.

regions (as the graph moves to the right), baseline levels of rainfall are already adequate, so variations in rainfall have much less impact.

Much of this is perhaps not surprising. What is less apparent is whether such rainfall shocks have had any impact on the expansion of cropland over the past few decades.

FIGURE 2.1 **Impact of Rainfall Shocks on Crop Productivity, by Rainfall Quantile**

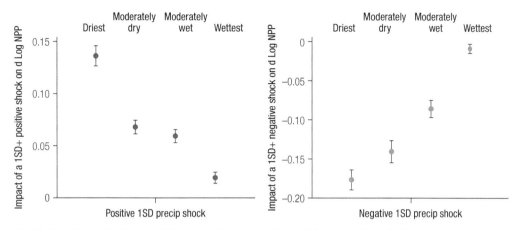

Note: This figure shows point estimates and 95 percent confidence intervals from coefficients obtained for each of the four quantiles, representing grid cells with lowest amounts of long-term average rainfall, to those with the highest amounts of long-term average rainfall. In each graph, plotted coefficients capturing the change in log of NPP are estimated using separate regressions, but negative and positive coefficients are estimated jointly. 1SD = 1 standard deviation; NPP = net primary productivity.

Dry Shocks Can Accelerate Cropland Expansion

Over the past decade, the world has lost 2.3 million square kilometers of forested land. It is estimated that 80 percent of this loss is a consequence of the ever-expanding agricultural frontier (Hansen et al. 2013). The global competition for agricultural land and forest resources remains central to policy discussions striving for a food-secure and low-carbon future. It is also at the heart of what is known as the *Borlaug Hypothesis*—the assertion that a rise in agricultural productivity may lead to *less* farmland expansion and therefore less deforestation. The idea behind this hypothesis is that if yields increase, farmers will have less need to expand cropland. If Borlaug's conjecture holds, this would suggest that rainfall-induced reductions in crop yields are responsible for farmland expansion and, as a consequence, a decline in forest cover.

But the opposite response could also occur. Much of the literature in agricultural economics assumes that farmers are "risk averse." With risk-averse behavior, rainfall uncertainty would encourage cautious farmers to lower their exposure to heightened risks and hence scale back the amount of land exposed to rainfall shocks. This could, for instance, occur through greater investments in smaller plots of land (intensification) as a defensive buffering strategy (Villoria, Byerlee, and Stevenson 2014). If risk aversion prevails, then adverse rainfall shocks would be associated with lower rates of cropland expansion.

Given the pace of global deforestation, it is important to determine which of these hypotheses hold and whether, and to what extent, these trends have been influenced by rainfall-related shocks. But there is surprisingly little evidence on this issue, despite its global significance.

The empirical analysis indicates that dry shocks lead to a substantial expansion of cropland, consistent with the Borlaug hypothesis. In general, in the years following dry shocks, cropland expands significantly; and the drier the shocks that occurred over the past decade, the faster is this rate of expansion. In fact, dry rainfall shocks account for about 60 percent of the rate of expansion in cropland over the past three decades. In contrast, wet rainfall shocks are less problematic, and there is no clear evidence that they induce changes in land-use patterns. The possible economic drivers behind such asymmetric responses to wet and dry shocks are explained in greater detail in box 2.2.

Current cropland expansion rates are unsustainable in many regions, and land clearing, which is responsible for about 6–17 percent of CO_2 emissions caused by humans, is already one of the larger contributors to CO_2 emissions (Baccini et al. 2012). Thus, any further cropland expansion could have far-reaching consequences for both agricultural production and climate stability, leading potentially to a vicious cycle of land clearing and increased vulnerability to rainfall fluctuations. There are other localized risks to agriculture, too. The volume of water that is available for farming and other uses typically scales proportionately with the size of the watershed that is left intact. When forests are converted into cropland, water storage potential declines and soil erosion accelerates. Thus, protecting forests in watersheds not only provides more viable sources of water for

---

BOX 2.2. **A "Safety-First" Response to Rainfall Shocks**

What might cause farmers to expand their cropland in the face of water deficits? One plausible explanation is that it is a *safety-first* adaptation technique. Repeated years of insufficient water endowments, and the depressed crop yields that result, send a signal to farmers that yields in future years may continue to remain low. A farmer who seeks a certain level of income, perhaps for subsistence, will have limited options for increasing total production. If yields are constantly depressed, the most obvious means to increase total production is to expand the area that is planted, even if this means increased exposure to rainfall risks.

Evidence that "safety first" may be the mechanism behind cropland expansion comes from splicing the data set. When only high-income countries are examined, where agriculture is highly commercialized and farmers are better equipped to adapt to water shortages (through irrigation, insurance, or savings), the effect (that is, cropland expansion as a coping mechanism for water deficits) disappears. But when the poorest countries— those classified as low income—are isolated, the effect more than doubles, with each year with a dry rainfall shock leading to a 0.15 percent increase in cropland expansion. These countries have the highest ratio of subsistence and low-technology farmers, and therefore the "safety first" hypothesis would suggest that poorer countries have a greater sensitivity to rainfall shocks.

agriculture at the local level but also buffers against climate change–driven variability that affects agricultural yields.

The vicious cycle between rainfall shocks, land clearing, and increased vulnerability to rainfall fluctuations becomes a tragic cycle when one considers the impacts on the livelihoods of the poor. It is well established (Hallegatte et al. 2015) that many poor rural families depend on forests and ecosystems for their income. In Latin America and Sub-Saharan Africa, forest- and environmental-derived income can account for as much as 30 percent of total income. Furthermore, when households face a shock, the income derived from forests is used as a safety net to meet basic needs. By depleting this source of income after periods of dry shocks, rural farmers are also eating away at this important safety net and their natural capital, reducing their resilience to future rainfall shocks.

The link between rainfall shocks and increased deforestation identified in this book is not a mere statistical artifact of the data. A more detailed assessment in Madagascar—a fragile biome—which is described in annex 2A, confirms this broad pattern using fine-scale forest cover data. When a dry shock occurs during the growing season in Madagascar, deforestation increases by 10–20 percent during the next planting season, suggesting that farmers respond by expanding their croplands at the expense of the forests. This implies that effectively protecting forests with environmental policies may be more difficult following abnormally dry seasons. This also suggests the need for policies that buffer vulnerable households from the consequences of variable rainfall. Safety nets that buffer households against drought may indeed be more effective at preventing deforestation than zoning and forest protection policies.

Implications for Food Security

Rainfall-induced changes in agricultural production have implications for global and local food security. Every year, the world loses a sizable portion of its food production to dry shocks. Between 2001 and 2013, enough calories to feed 81 million people every day were lost to dry shocks each year. This is sufficient food to feed an entire country the size of Germany or Turkey. Specifically, total losses in areas that experienced dry shocks amounted to an average annual reduction of 59.2 trillion kilocalories (kcals) over that time period.[5] Losses of this magnitude undermine food security in ways that are unevenly distributed around the globe. As shown in map 2.1, regions that suffered large declines in production because of rainfall shocks include southern Mexico and Central America, northern South America, Western Europe, most of the Sahel and Southern Africa, Indonesia, and southern Australia.

On the other hand, wet shocks did lead to an annual average net gain of 45.3 trillion kcals per year in other regions. The biggest winners were in the Midwest of the United States, southeastern South America, Eastern Europe, and northern coastal China. In India, as well as in parts of Brazil,

MAP 2.1 **Average Annual Gains and Losses in Food Production as a Result of Rainfall Shocks, 2001–13**

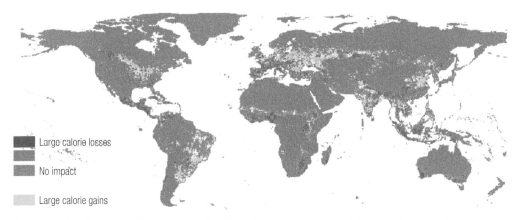

Large calorie losses

No impact

Large calorie gains

Sources: NPP data are from the annual MOD17A3 measures from 2000 to 2013 generated by the Numerical Terradynamic Simulation Group (NTSG) at the University of Montana (Zhao et al. 2005). Estimates shown in the map are based on World Bank calculations.

MAP 2.2 **Food Production Losses over the Past Decade and Food Insecurity, 2001–13**

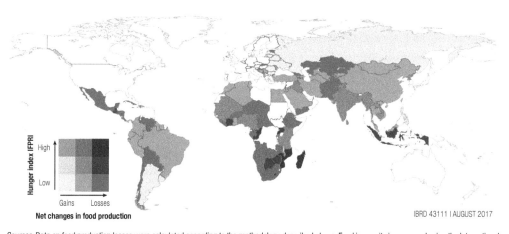

Hunger index IFPRI

High

Low

Gains Losses

Net changes in food production

IBRD 43111 | AUGUST 2017

Sources: Data on food production losses were calculated according to the methodology described above. Food insecurity is measured using the International Food Policy Research Institute (IFPRI) Hunger Index. Countries that are excluded (shown in white) either have no data for the Hunger Index or are not surveyed.

impacts were highly localized, with areas of gains and losses overlapping. It is important to note also that these averages mask significant annual fluctuations. A bountiful harvest in one year cannot always buffer against losses in the next year. Nor can production gains in one part of the world always offset losses in another, despite freer trade. These estimates therefore likely underestimate the true impact year-to-year.

To identify the most vulnerable hotspots, country-level food shortages are overlaid with a map of food security. This is displayed in map 2.2, which shows the biggest hotspots of vulnerability—these countries have lost significant percentages of their food production and are also deemed food-insecure under the Hunger Index of the International Food Policy Research Institute (IFPRI). Several stand out as being especially fragile, notably Botswana, Madagascar, Mozambique, and Zimbabwe. These are the countries where large food production losses as a result of rainfall shocks occurred against a backdrop of food insecurity, heightening the impact.

The Role of Infrastructure as a Buffer against the Effects of Rainfall Variability

Confronted with fickle rainfall, humans have impounded rivers and transported water over long distances using canals, pipes, and aqueducts since at least Roman times. Surprisingly little has changed, and these basic supply-side approaches still remain the preferred way to insulate economies against fluctuations in rainfall. With rainfall variability projected to increase throughout the 21st century, relatively modest losses in food production today could turn into severe losses in the not-too-distant future. So it is imperative to determine the effectiveness and robustness of such solutions as a protective buffer against more extreme rainfall.

The statistical evidence presented in volume 2 suggests that, in most cases, the reliance on infrastructure is indeed justified. On average, large irrigation infrastructure provides a healthy boost to NPP, the primary metric for crop production used in this analysis. The benefits are largest in middle-income countries, where NPP increases by 12 percent on average in regions where infrastructure is present.

Perhaps more important, in most areas that are equipped for irrigation, NPP shows little sensitivity to rainfall variability, both for wet and dry shocks. This implies that irrigation infrastructure provides a complete buffer against rainfall shocks in these areas. And since crop yields are protected from rainfall shocks, there is no expansion of cropland when farms are equipped with irrigation. The implication is that policies and investments that insulate agricultural yields deliver simultaneous benefits both to farms and to forests.

But there are notable exceptions. In some arid areas, as well as in low-income countries, large irrigation infrastructure is found to be less effective at buffering agricultural yields against shocks and halting the expansion of cropland. In fact, there are regions where the presence of irrigation infrastructure accentuates the impact of shocks on agricultural yields. In these areas, NPP declines when dry shocks occur; and if irrigation infrastructure is in place, NPP declines even further.

This finding may seem surprising at first, but it is also entirely predictable. Infrastructure fails to buffer against dry shocks in those arid regions where a disproportionate amount of land is used to cultivate water-intensive crops—such as rice, sugarcane, cotton, and maize. The ironclad laws of supply and demand dictate that when a useful resource is provided for free (or almost free), it will inevitably be consumed in its entirety. Hence, the supply of free or underpriced water provided by irrigation services in arid areas buoys the cultivation of water-intensive crops, which increases vulnerability to drought, which in turn magnifies the impacts of dry shocks. The end result is *maladaptation*—a dependence on water-sensitive crops in arid areas that increases vulnerability to dry shocks.

Other studies have documented a similar maladaptive reaction to irrigation in specific regions. For instance, on the American Great Plains in the second half of the 20th century, access to irrigation from the Ogallala aquifer led farmers to grow water-intensive crops, enhancing drought sensitivity (Hornbeck and Keskin 2014). And there are historical precedents too, with the decline of the Mayan communities associated with a similar pattern of overreliance on irrigated, drought-sensitive crops.[6] This book demonstrates that this *paradox of supply* is far too common a problem in areas where water is naturally scarce and its demand is uncontrolled.

Coupling Investments with Policies Can Overcome the Paradox of Supply

The implications of these findings for stabilizing food production are significant. Investments and supply-side solutions are essential to enhance productivity in agriculture. But the inescapable laws of supply and demand imply that to be effective, they also must be coupled with demand-side incentives to build resilience to the inevitable intensifying incidence of rainfall shocks.

Pricing mechanisms that shift the cost of water closer to its true economic value can go a long way toward incentivizing sustainable water practices in agriculture. The use of water rights and water permits, which bestow fixed, yet tradable proportions of water to users, is another instrument that can help realize the economic value of water, while also incentivizing its prudent and waste-conscious use. These issues are investigated in greater detail in chapter 5.

While irrigation can provide a good buffer against rainfall variability and bolster crop yields, it is not a prudent response to water scarcity everywhere. The best and most productive sites to build infrastructure have been exhausted in many parts of the world, and most of the world's river basins are already highly stressed. Indeed, as many as 4 billion people already live in regions that suffer severe water stress during at least part of the year (Mekonnen and Hoekstra 2016)—where net surface and groundwater withdrawals exceed available supply. These basins, shown in map 2.3, are effectively "closed," and the addition of new irrigation infrastructure becomes a zero-sum game, where new users receive water at the expense of current users.

Better management of water is therefore going to be vital for meeting the world's future demands. Converting much of the world's food production into *climate smart agriculture* (CSA) could go a long way toward reducing water wastage, overuse, and pollution. CSA combines smart

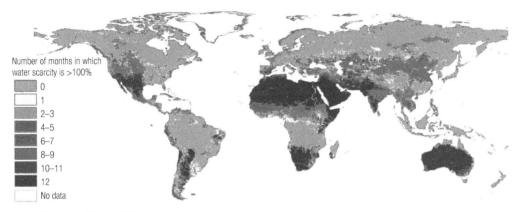

MAP 2.3 **Regions That Experience Severe Water Scarcity throughout the Year, 1996–2005**

Number of months in which
water scarcity is >100%

- 0
- 1
- 2–3
- 4–5
- 6–7
- 8–9
- 10–11
- 12
- No data

Source: Mekonnen and Hoekstra 2016.
Note: Water scarcity > 100 percent reflects a situation where net surface and groundwater withdrawals plus environmental flow requirements are greater than the available supply.

policies, financing, and technologies to achieve a "triple win"—increased agricultural productivity, enhanced climate resilience of crops, and reduced greenhouse gas emissions from agricultural production (World Bank and International Energy Agency 2015). New supplies, technologies, efficiency enhancements, and demand management will all be needed, as discussed in chapter 5.

Concluding Comments

Rainfall variability has already had profound impacts on the agricultural sector that are yet to be fully understood. With climate change expected to amplify these impacts, the global food industry will have to adapt to increased uncertainty and growing water scarcity to feed an ever-growing population. Better policies and smarter investments will be needed to overcome these unprecedented challenges. Failure to adapt will have far-reaching consequences for global food security, as well as for the livelihoods of those who depend upon farm incomes. The next chapter examines the impacts that rainfall shocks have on rural households.

Annex 2A: Rainfall Shocks, Farmers, and Forests: Weather-Induced Deforestation in Madagascar

Over 1 billion people depend on access to forests for their livelihoods (World Resources Institute 2005). The collection of timber and nontimber forest products accounts for almost a third of their incomes, a share often comparable to agriculture (Anglesen et al. 2014). Yet, because of population growth, changes in food consumption habits, and a limited capacity of many nations to enforce laws on their territories, the world has lost more than 7.5 percent of its forests between 2000 and today.

In addition to these traditional factors that cause deforestation, this annex highlights how rainfall variability can be at the source of less (or even un-) documented weather-induced deforestation. Indeed, because rainfall shocks alter agricultural yields and because farmers adapt the size of their croplands in the aftermath of shocks, deforestation is itself sensitive to rainfall shocks. Evidence of this is presented here using fine-scale deforestation data from Madagascar.

The Madagascar Context

Madagascar is one of the least-developed economies on the planet, with 78 percent of the population living on less than US$1.9 a day. Eighty percent of the workforce is in the agricultural sector, and of the 25 million inhabitants of the country, approximately 1 million live at the edges of natural forests (Desbureaux 2016). These farmers rely on clearing forests to access new arable lands for rain-fed agriculture. Since 1950, 50 percent of Madagascar's forests have disappeared (figure 2A.1) (Harper et al. 2007). And this rate has not slowed down; in the year 2014 alone, close to 2 percent of remaining forests was cleared (Petersen et al. 2015).

The richness of Madagascar's forests is unique. The vast majority of the country's diverse, but threatened, biodiversity is endemic to the island. The emblematic lemur is a perfect illustration of this situation, as 94 of the 101 endemic lemur species are threatened with extinction (IUCN 2014). Dense rainforests store large amounts of carbon and protect soil against erosion, which helps preserve the productivity of agricultural lands and future water provision. The water-capturing and storage services that the forests provide strengthen river flows and increase the capacity of energy production, which is particularly important in Madagascar, where hydropower accounts for 60 percent of total electricity generation.

Empirical Analysis

In order to assess whether rainfall shocks affect deforestation, fine-scale satellite images were compiled to track deforestation annually between 2000 and 2013 at a 30 meter x 30 meter scale. These data were synthetized in a 0.1-degree grid and merged with the same weather data described in box 2.1. Regression analysis was then performed, following a methodology similar to the one described in box 2.1, with the outcome variable being a satellite-based measure of deforestation.

Results indicate that during a year with an abnormally low level of rainfall, deforestation in Madagascar is 9.5 percent higher than during a year with average rainfall. When rainfall is abnormally high, overall deforestation is 13.5 percent lower. Consistent with an agriculture story, shocks that occur during the growing seasons have the biggest impacts. In addition, dry shocks lead to twice as much deforestation when they occur in areas where the agricultural frontiers are already currently expanding,

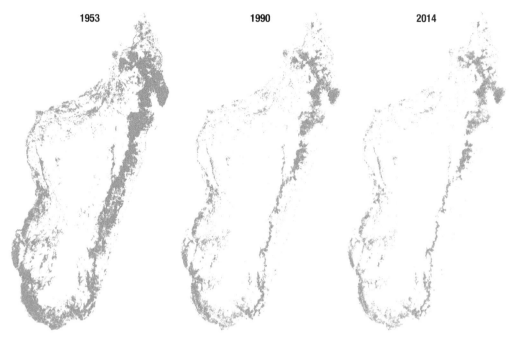

Source: BioSceneMada (https://bioscenemada.cirad.fr/maps/).

implying that local communities are nearby.[7] Inside these forests, deforestation is 22 percent higher when the year is abnormally dry as compared to a normal year. However, wet shocks have no impact on deforestation. These results largely mirror the cropland results shown in chapter 2. They also echo the safety-first strategy of coping with risks, described in box 2.2.

Weather-Induced Deforestation Makes the Conservation of Forests More Difficult

Madagascar is particularly exposed to rainfall shocks and will be even more exposed in the future, with climate change (Dewar and Richard 2007). Extreme events, such as floods and droughts, have become more frequent and more intense over the past two decades. Recent years, for example, have been marked both by a severe drought throughout the country, and Hurricane Enawo, which flooded much of the lowlands in the eastern part of the island. Dry shocks in Madagascar were 50 percent more frequent relative to the rest of the world between 2000 and 2013.[8]

The effectiveness of environmental protection policies has been a central debate in the global political agenda since the early 1990s. In many tropical countries, academic evidence suggests that the decrease in deforestation brought on by conservation and sustainable policies has often been limited (Baylis et al. 2016; Eklund et al. 2016; Geldman et al. 2013).

In Madagascar, an important network of protected areas covers 10 percent of the territory. However, this analysis has also found that the increase in deforestation during abnormally dry years is similar inside and outside protected areas. This calls into question the effectiveness of forest management policies and suggests that deforestation resulting from climate variation may be more difficult to prevent.[9]

When droughts occur, farmers suffer. And when farmers suffer, forest cover declines, and achieving sustainable objectives becomes more challenging. This annex presents early work on these issues, and future research will be critical for understanding the full picture.

Notes

1. Rangarajan 1992, as quoted in Barnett 2016.
2. See Dell et al. 2012 for a survey of the academic literature.
3. See volume 2 for a fuller discussion: www.worldbank.org/UnchartedWaters.
4. A data set satisfies statistical stationarity when its mean, variance, and other statistical properties are constant over time. In panel or time series analysis, this property is important for generating unbiased estimates.
5. This assumes a 2,000 kcals per day, per person, or 730,000 kcals per person, per year.
6. A recent article by Kuil et al. (2016) suggests that the decline of the Mayan civilization in the 9th century might have been exacerbated by overreliance on artificial reservoirs. Although the reservoirs assisted in capturing water used for irrigation, and allowed population densities to increase, when rainfall levels declined and the reservoirs dried up, the large populations that they supported were put at risk.
7. We identify these forests using the spatial census of community forests conducted by Lohanivo (2013).
8. In the climate data used here, there was a 20 percent probability of a dry shock occurring in any gridcell/year combination in Madagascar from 2000 and 2013. This same probability was 14 percent worldwide.
9. When estimating the impact of conservation policies, such as Protected Areas, one has to take into account the nonrandomness of their location (Joppa and Pfaff 2009). We use entropy-balancing matching techniques (Hainmueller 2012) to control for location bias before performing regressions.

References

Abdi, A. M., J. Seaquist, D. E. Tenenbaum, L. Eklundh, and J. Ardö. 2014. "The Supply and Demand of Net Primary Production in the Sahel." *Environmental Research Letters* 9 (9): 094003.

Angelsen, A., P. Jagger, R. Babigumira, B. Belcher, N. Hogarth, S. Bauch, J. Börner, C. Smith-Hall, and S. Wunder. 2014. "Environmental Income and Rural Livelihoods: A Global-Comparative Analysis." *World Development* 64 (S1): S12–S28.

Auffhammer, M., V. Ramanathan, and J. R. Vincent. 2006. "Integrated Model Shows That Atmospheric Brown Clouds and Greenhouse Gases Have Reduced Rice Harvests in India." *Proceedings of the National Academy of Sciences* 103 (52): 19668–72.

A. Baccini, S. J. Goetz, W. S. Walker, N. T. Laporte, M. Sun, D. Sulla-Menashe, J. Hackler, P. S. A. Beck, R. Dubayah, M. A. Friedl, S. Samanta, and R. A. Houghton. 2012.

"Estimated Carbon Dioxide Emissions from Tropical Deforestation Improved by Carbon-Density Maps." *Nature Climate Change* 2 (3): 182–85.

Barnett, C. 2016. *Rain: A Natural and Cultural History.* New York: Broadway Books.

Baylis, K., J. Honey-Rosés, J. Börner, E. Corbera, D. Ezzine-de-Blas, P. J. Ferraro, R. Lapeyre, U. M. Persson, A. Pfaff, and S. Wunder. 2016. "Mainstreaming Impact Evaluation in Nature Conservation." *Conservation Letters* 9: 58–64.

Blanc, E., and E. Strobl. 2013. "The Impact of Climate Change on Cropland Productivity: Evidence from Satellite Based Products at the River Basin Scale in Africa." *Climatic Change* 117 (4): 873–90.

Burke, M., S. M. Hsiang, and E. Miguel. 2015. "Global Non-Linear Effect of Temperature on Economic Production." *Nature* 527: 235–39.

Dell, M., B. F. Jones, and B. A. Olken, 2012. "Temperature Shocks and Economic Growth: Evidence from the Last Half Century." *American Economic Journal: Macroeconomics* 4 (3): 66–95.

Desbureaux, S. 2016. "Efficacité des Politiques de Luttes Contre la Déforestation et Logiques d'Action Collective à Madagascar." ["Effectiveness of Policies at Reducing Deforestation and the Logic of Collective Action in Madagascar"]. Doctoral Thesis, University of Auvergne.

Dewar, R., and A. Richard. 2007. "Evolution in the Hypervariable Environment of Madagascar." *Proceedings of the National Academy of Sciences* 104 (34): 13723–27.

Eklund, J., F. G. Blanchet, J. Nyman, R. Rocha, T. Virtanen, and M. Cabeza. 2016. "Contrasting Spatial and Temporal Trends of Protected Area Effectiveness in Mitigating Deforestation in Madagascar." *Biological Conservation* 203: 290–97.

Fishman, R. 2016. "More Uneven Distributions Overturn Benefits of Higher Precipitation for Crop Yields." *Environmental Research Letters* 11 (2): 024004.

Geldmann, J., M. Barnes, L. Coad, I. D. Craigie, M. Hockings, and N. D. Burgess. 2013. "Effectiveness of Terrestrial Protected Areas in Maintaining Biodiversity and Reducing Habitat Loss: Systematic Review." *Collaboration for Environmental Evidence* 161 (May): 230–38.

Hainmueller, J. 2012. "Entropy Balancing for Causal Effects: A Multivariate Reweighting Method to Produce Balanced Samples in Observational Studies." *Political Analysis* 20 (1): 25–46.

Hallegatte, S., M. Fay, M. Bangalore, T. Kane, and L. Bonzanigo. 2015. *Shock Waves: Managing the Impacts of Climate Change on Poverty.* Washington, DC: World Bank.

Hansen, M. C., P. V. Potapov, R. Moore, M. Hancher, S. A. Turubanova, A. Tyukavina, D. Thau, S. V. Stehman, S. J. Goetz, T. R. Loveland, A. Kommareddy, A. Egorov, L. Chini, C. O. Justice, and J. R. G. Townshend. 2013. "High-Resolution Global Maps of 21st-Century Forest Cover Change." *Science* 342 (6160): 850–53.

Harper, G. J., M. K. Steininger, C. J. Tucker, D. Juhn, and F. Hawkins. 2007. "Fifty Years of Deforestation and Forest Fragmentation in Madagascar." *Environmental Conservation* 34 (4): 325–33.

Hornbeck, R., and P. Keskin. 2014. The Historically Evolving Impact of the Ogallala Aquifer: Agricultural Adaptation to Groundwater and Drought." *American Economic Journal: Applied Economics* 6 (1): 190–219.

IUCN (International Union for Conservation of Nature). 2014. *The IUCN Red List of Threatened Species.* Version 2014.1. IUCN, Cambridge. http://www.iucnredlist.org.

Joppa, L. N., and A. Pfaff. 2009. "High and Far: Biases in the Location of Protected Areas." *PLoS One* 4 (12): e8273.

Kuil, L., G. Carr, A. Viglione, A. Prskawetz, and G. Blöschl. 2016. "Conceptualizing Socio-Hydrological Drought Processes: The Case of the Maya Collapse." *Water Resources Research* 52 (8): 6222–42.

Lobell, D. B., J. A. Hicke, G. P. Asner, C. B. Field, C. J. Tucker, and S. O. Los. 2002. "Satellite Estimates of Productivity and Light Use Efficiency in United States Agriculture, 1982–98." *Global Change Biology* 8 (8): 722–35.

Lobell, D. B., G. L. Hammer, G. McLean, C. Messina, M. J. Roberts, and W. Schlenker. 2013. "The Critical Role of Extreme Heat for Maize Production in the United States." *Nature Climate Change* 3 (5): 497–501.

Lohanivo, A. 2013. "Evaluation quantitative de la mise en oeuvre de la loi GELOSE: Recensement des TG dans 13 Régions de Madagascar" ["Quantitative Evaluation of the Implementation of the GELOSE Law: Census of Community-Managed Natural Resources in 13 Regions of Madagascar"]. In *Rôle et place des transferts de gestion des ressources naturelles renouvelables dans les politiques forestières actuelles à Madagascar [The Role of Place of Transferring the Management of Renewable Natural Resources in Current Forestry Policies in Madagascar].* France.

Lu, X., and Q. Zhuang. 2010. "Evaluating Climate Impacts on Carbon Balance of the Terrestrial Ecosystems in the Midwest of the United States with a Process-Based Ecosystem Model." *Mitigation and Adaptation Strategies for Global Change* 15 (5): 467–87.

Mekonnen, M., and A. Hoekstra. 2016. "Four Billion People Facing Severe Water Scarcity." *Science Advances* 2 (2): e1500323.

Naylor, R. L., W. P. Falcon, D. Rochberg, and N. Wada. 2001. "Using El Niño/Southern Oscillation Climate Data to Predict Rice Production in Indonesia." *Climate Change* 50: 255–65.

Petersen, R., N. Sizer, M. Hansen, P. Potapov, and D. Thau. 2015. "Satellites Uncover 5 Surprising Hotspots for Tree Cover Loss." Blog post, September 2. World Resources Institute, Washington, DC. http://www.wri.org/blog/2015/09/satellites-uncover-5 -surprising-hotspots-tree-cover-loss.

Rangarajan, L. N., ed. 1992. *The Arthashastra.* Haryana, India: Penguin Books.

Running, S. W., R. R. Nemani, F. A. Heinsch, M. Zhao, M. Reeves, and H. Hashimoto. 2004. "A Continuous Satellite-Derived Measure of Global Terrestrial Primary Production." *Bioscience* 54 (6): 547–60.

Sawano, S., T. Hasegawa, S. Goto, P. Konghakote, A. Polthanee, Y. Ishigooka, T. Kuwagata, and H. Toritani. 2008. "Modeling the Dependence of the Crop Calendar for Rain-Fed Rice on Precipitation in Northeast Thailand." *Paddy Water Environment* 6: 83–90.

Tum, M., and K. P. Günther. 2011. "Validating Modelled NPP using Statistical Yield Data." *Biomass and Bioenergy* 35 (11): 4665–74.

Verón, S. R., D. de Abelleyra, and D. B. Lobell. 2015. "Impacts of Precipitation and Temperature on Crop Yields in the Pampas." *Climatic Change* 130 (2): 235–45.

Villoria, N. B., D. Byerlee, and J. Stevenson. 2014. "The Effects of Agricultural Technological Progress on Deforestation: What Do We Really Know?" *Applied Economic Perspectives and Policy* 36 (2): 211–37.

World Bank and International Energy Agency. 2015. *Progress Toward Sustainable Energy for All 2015.* Washington, DC: World Bank.

World Resources Institute. 2005. *The Wealth of the Poor: Managing Ecosystems to Fight Poverty.* Washington, DC: World Resources Institute.

Zhao, M., F. A. Heinsch, R. R. Nemani, and S. W. Running. 2005. "Improvements of the MODIS Terrestrial Gross and Net Primary Production Global Data Set." *Remote Sensing of Environment* 95 (2): 164–76.

3

When Rainfall Is Destiny: The Long-Lasting Impacts of Water Shocks on Families

There is always one moment in childhood when the door opens and lets the future in.

—Graham Greene, *The Power and the Glory*

Key Chapter Findings

- Rainfall shocks experienced in a child's earliest years can have long-lasting impacts that trap households in poverty.
- Females born in rural areas around the time of exceptionally dry spells are less wealthy as adults, have lower levels of human capital, and are less empowered compared to those born during periods of normal rainfall.
- In contrast, females born during exceptionally wet spells are wealthier and have lower fertility rates as adults.
- The negative impacts of rainfall deficits can traverse generations.

Water shocks can have disastrous contemporaneous impacts. Droughts can lead to food crises, economically stifle farmers, and set off epidemics, while floods can inundate cities, wipe away lives and livelihoods, and destroy critical infrastructure. Policy responses tend to focus on the immediate visible impacts and, as such, governments and the international community view these events as short-term traumas that vanish once the rain starts falling again or the floodwaters recede. And indeed, after an urban flood occurs, it is often true that aggregate economic activity in affected areas returns to normal levels within a year of the flood (Kocornik-Mina et al. 2015). Largely ignored, however, are the long-term consequences of such disasters for the individuals who live through them. The effects may be undetectable in the short term, but can follow these people throughout their entire lives.

This chapter examines the long-term consequences of water shocks and documents a dismal set of statistics that has profound consequences for human development. It finds that in rural Africa, a rainfall shock experienced around the time of a child's birth can impair that child's development, her educational performance, and her wealth as an adult, and even harm the health of her children. The significance of these findings and their ramifications for development policy cannot be overstated. They demonstrate that rainfall can truly shape destinies, and that our current understanding of the impact of rainfall on society may be severely limited.

The Importance of Early-Life Environmental Conditions

The results reported in this chapter are related to a substantial body of evidence on the *fetal origins hypothesis*, which asserts that the conditions that a child experiences early in life can have significant and often irreversible long-term consequences.[1] Research in this field has shown that stresses resulting from a variety of events, including epidemics, natural disasters, wars, and climate extremes, can interrupt the physical and mental development process of a fetus or infant, and lead to long-term negative consequences. These include poorer health and height attainment, reduced cognitive abilities and educational achievement, poorer labor market outcomes, and reductions in household income and wealth.[2] There are numerous channels through which adverse conditions in infancy can have long-term consequences, but health, hygiene, and nutritional intake appear to be among the most important. For instance, poor diet and insufficient calorie intake by pregnant women and young children can harm physiological development, with notably large impacts on brain development.[3] Beyond this, they may also affect adult outcomes through other channels, such as altered schooling decisions and socioeconomic status. As a result, children who experience adverse conditions in their earliest years may never recover. This relationship is summarized in figure 3.1. Furthermore, there is evidence to suggest that females may suffer disproportionately large long-term consequences of adverse environmental conditions.[4]

Sub-Saharan Africa Is a Region Highly Vulnerable to Rainfall Shocks

The focus of the analysis in this chapter is Sub-Saharan Africa, where agriculture is the largest source of income for 90 percent of the rural population and there is high reliance on rain-fed agriculture. In fact, less than 4 percent of arable land in Africa is irrigated (UNEP 2010) and as a result

FIGURE 3.1 **Channels through which Rainfall Shocks Can Have Long-Term Impacts**

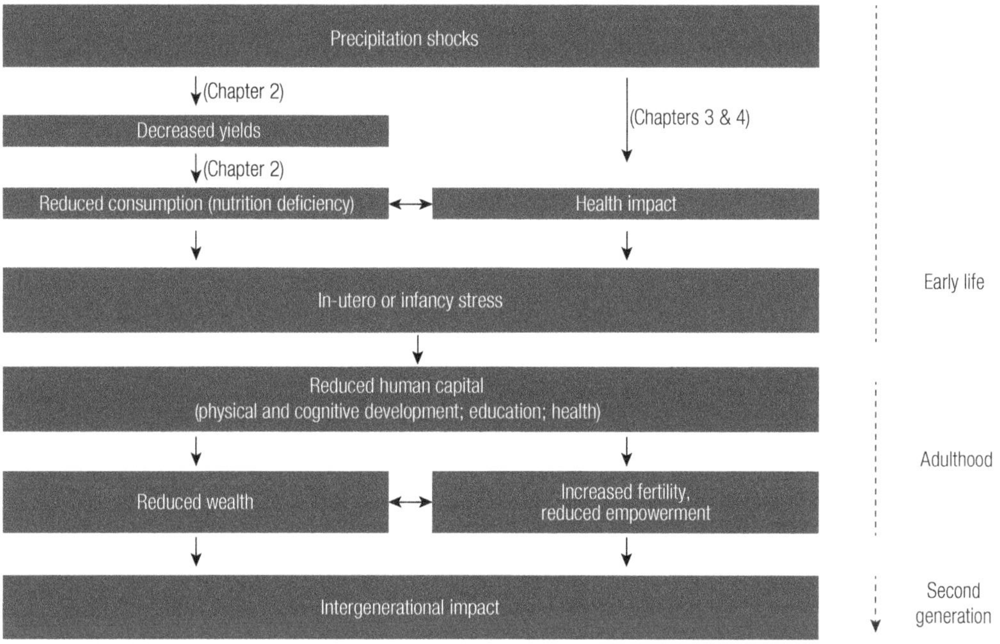

Source: Adapted from Carrillo, Fishman, and Russ 2015.

the region is acutely vulnerable to rainfall-related shocks. In many parts of Sub-Saharan Africa, the poor are even more exposed to weather extremes than their wealthier counterparts. They live on more marginal lands, which are more prone to both flooding and droughts. And when these floods and droughts do occur, the poor are more vulnerable to their impacts, typically losing a larger share of their wealth. They have less infrastructure to protect them from natural hazards, and their property is of lower quality, and thus is more likely to be damaged (Hallegatte et al. 2015).

The analysis in this chapter is based on data from the Demographic and Health Surveys (DHS) in 19 Sub-Saharan African countries (see box 3.1).[5] These data sets were used to investigate the relationship between a rainfall shock occurring during a child's infancy and a range of adult outcomes including wealth, education, and fertility, as well as second-generation outcomes. The analysis is undertaken over wide geographic areas and long periods of time, implying that the uncovered relationships are robust, and not merely some artifact of the chosen context.

Approximately 72 percent of respondents (representing just under 77,000 women) in the data set were born and raised in rural households. Of these rural residents, 55 percent of them live in households that are in the bottom two wealth quintiles, as defined by the DHS wealth index.

Study Locations, Data, and Methodology

In order to measure the long-term impacts of rainfall shocks, data from the Demographic and Health Surveys (DHS)[a] are used. The DHS are large, nationally representative surveys that cover more than 90 countries worldwide. With a particular focus on women, they address a range of issues related to population and health. Map B3.1 displays the countries included in the study, with black dots showing approximate locations of the communities where the households live. This chapter focuses on those Sub-Saharan African countries where there were adequate data on location of birth, indicators measuring wealth, and GPS coordinates.

The methodology employed here is similar to that used in chapter 2, described in box 2.1. However, this chapter focuses on the impacts of *large* rainfall shocks—specifically, situations in which annual rainfall is 2 or more standard deviations above (large wet shock) or below (large dry shock) its long-run mean. While small (1 standard deviation) shocks are also found to have an impact, their effects are generally smaller.

MAP B3.1 **Countries Studied**

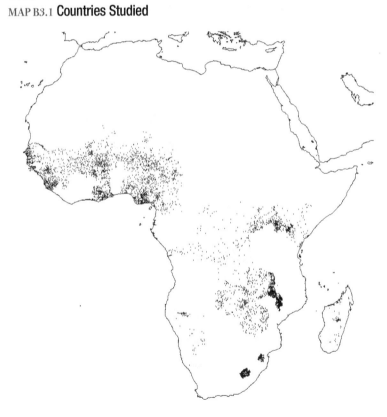

a. http://www.dhsprogram.com/.

These women have, on average, completed a mere three and a half years of schooling, and their average height is 157 centimeters (less than 5.2 feet). The average mother in the sample has given birth to 4.3 children, and some of the young children of these women exhibit signs of malnutrition and suffer from stunting, wasting, or are underweight. These statistics serve to underscore the fact that the women who are the focus of this chapter represent a highly vulnerable population.

The Impacts of Rainfall Shocks in Infancy Persist Well into Adulthood

The empirical analysis strongly suggests that rainfall shocks experienced by females as infants have durable, long-term impacts that stretch well into adulthood. A woman living in a rural area that experienced a *large* dry shock during her infancy is found to have less wealth as an adult compared to women in similar circumstances but who were born in years of normal rainfall. Conversely, favorable weather conditions (large wet shocks) in infancy can have a positive impact on adult wealth. Adult household wealth tends to be about 8 percent lower for women who experienced dry shocks in infancy, relative to those that did not. Women who lived through wet shocks have 7 percent higher levels of wealth than those who did not experience a shock. Surprisingly, irrigation systems do not seem to buffer these vulnerable communities from the shocks (box 3.2), perhaps because baseline levels of poverty and vulnerability are already so high that they

---

BOX 3.2. **The Buffering Ability of Irrigation Infrastructure**

Data on the location of large-scale irrigation infrastructure were used to analyze whether irrigation infrastructure can help buffer against rainfall variability and prevent the long-term consequences of rainfall shocks. The results are somewhat ambiguous. While irrigation infrastructure does tend to increase quality of life, and women born in areas equipped with it do tend to become wealthier adults relative to those born in other areas, there is little evidence that irrigation infrastructure has been effective at buffering against the type of rainfall shocks studied here, even after accounting for the nonrandom nature of infrastructure placement. There are a number of possible reasons for this finding.

First, the births studied here occurred as early as the 1950s and ended in the 1990s. Over that time, in the countries studied, irrigation infrastructure was quite rare. This makes it difficult to find evidence of either a meaningful impact or a lack of one. Furthermore, chapter 2 illustrated that the ability of infrastructure to mitigate the impact of rainfall shocks on agricultural output is limited in poor and arid countries, where there is maladaptation. Moreover, rainfall shocks may have direct impacts beyond those on agricultural output (such as those related to disease and environment), against which irrigation infrastructure is not designed to offer protection. For instance, the presence of irrigation infrastructure may have no impact on health and sanitation channels, which may be important.

are insufficient to insulate communities, suggesting the need for alternative interventions.

Reasons for the effect of early-life rainfall shocks on adult wealth can be gleaned from the relationship between these shocks and key indicators of human capital, such as educational attainment and proxies for health. Rural women who experience dry shocks in their earliest years have, by the time they reach adulthood, significantly less formal education. The magnitude of this effect is equivalent to a 6 percent decline in schooling for the average woman in the sample.[6] Indeed, the results show that a woman who experienced a large dry shock in her infancy is 36 percent more likely to have fewer than two years of formal education.

Not only are these women less educated, but they are also about 0.6 centimeter shorter than women of similar circumstances born in years of normal rainfall. This is a widespread problem in Sub-Saharan Africa, where more than 35 percent of children under the age of five are considered stunted (more than 2 standard deviations below the reference height for age of their cohort) (World Development Indicators 2015).[7] Over the past century, health and nutritional improvements, while substantial, have only managed to increase the average female adult height by approximately 2 centimeters in Sub-Saharan Africa (NCD Risk Factor Collaboration 2016). Thus, a large dry shock in infancy can wipe out almost one-third of this gain. Stunting is of particular importance because it is typically a physical marker of cognitive impairment resulting from deprivations in early life. Perhaps as a result, height has repeatedly been shown to be a significant predictor of earnings (see box 3.3).

BOX 3.3. **The Economic, Social, and Physiological Importance of Height**

The use of adult height as an indicator of health may seem unusual. After all, in most contexts a taller height is seen as more of a genetic consequence than something related to environmental factors. And indeed, in the developed world that is usually true, with approximately 80 percent of variation in height being attributable to genetics (Silventoinen 2003).

However, throughout the developing world, and particularly in the poorest communities like the ones studied here, environmental factors can play a large role in influencing height. Adult height represents cumulative growth through infancy, childhood, and adolescence, and growth during any of these stages of development can be impacted by environmental factors (Institute of Medicine 2001; Resnik 2002). The important environmental factors in determining adult height are the quality of the uterine environment, and, after birth, the nutritional status and disease environment of childhood. Indeed, the first 1,000 days of life are a crucial period in determining growth (Prendergast and Humphrey 2014). Stunting, the condition of having height that is well below one's genetic potential, affects one-third of children under five years old in developing countries and is often related to bad sanitation facilities and frequent bouts of childhood diarrhea.

(continued on next page)

So why is height important? While it can indicate an early-life nutritional deficit, it may also be evidence of a latent loss in cognitive development. Because both physical and mental abilities are developing at the same time, a shortfall in one can signal a shortfall in the other. Further, there is a well-established relationship between height and economic success. Studies dating back over a century have found a clear link between professional achievements and adult height (Gowin 1915). There are several explanations for this. Genetics may play an important role, as studies of twins have shown that genes account for approximately 35 percent of the correlation between height and intelligence (Sundet et al. 2005). The remaining 65 percent is due to environmental factors. A small part of this portion may be explained by social factors such as self-esteem, social dominance, and discrimination.[a] The remainder is likely a mix of the link between physical development and cognitive development, as well as the additional productive efficiency that comes from being taller (and stronger). This last factor may be particularly important in the rural farming communities studied here.

a. See Case and Paxson (2008) and citations therein.

As figure 3.1 illustrates, there are multiple channels through which early life deprivations could encumber individuals. It is difficult to precisely identify the dominant pathway with the information available in the DHS. Stunting in this sample could either be a consequence of poor nutritional intake or ill health in infancy caused by an altered disease environment, from which the child never recovers (Martorell 1999).[8] There is suggestive evidence that nutrition and food availability could be a dominant factor in this context. Rural households are much more impacted by dry shocks than urban households, suggesting that low availability of food in subsistence farming households is a possible pathway, as suggested by the analysis in chapter 2, which demonstrates a strong relationship between rainfall shocks and agricultural productivity.

The Impacts of Large Fluctuations in Rainfall Persist into the Next Generation

If a rainfall shock in infancy is a woman's destiny, it may be that of her children, too. Disturbingly, shocks experienced by a mother in her infancy can be passed on to her offspring, with the effects showing up as lower health outcomes. Women who experienced a large dry shock in infancy are 29 percent more likely to have a child suffering from some form of anthropometric failure—that is, being significantly below average size in terms of height for age, or weight for age, or weight for height.[9] Given the array of adult outcomes that large fluctuations in early-life rainfall affect, this is perhaps unsurprising. But these findings add to the urgency of addressing the effects of adversity in infancy.[10]

The Cascading Impacts on the Gender Dimension

The long-term effects of rainfall shocks can even go beyond impacts related to health and economic outcomes. Varying rainfall during infancy plays a surprising role in reducing female agency and empowerment, and increasing fertility rates, perhaps through the effects of lower incomes or education. These findings reinforce the dismal conclusion that women may never recover from climate-induced shocks experienced in their earliest years, and that the effects also cascade into the behavioral domain.

Looking first at issues related to female empowerment, women who experienced a large dry shock in their infancy are found to be less likely to have a role in household decision making when they are adults. Other studies have noted that a woman's ability to exercise agency (that is, her ability to make decisions and act for herself) at any point in her life reflects conditions experienced in childhood (World Bank 2011). These empirical findings echo this fact: women who experienced a large dry shock in infancy are more likely to report that it is their husband/partner (or another person) who takes sole responsibility for decisions related to household purchases. Clearly this result is important for gender equality and also has wider consequences as a result of the impacts of gender equality on economic development (World Bank 2011).

Early-life rainfall shocks also affect fertility rates. Other types of shocks, such as famine or conflict, have been shown to impact fertility rates in ambiguous ways that are context specific. The results of this analysis show that women born around a period of above-average rainfall have fewer children overall, but there is no impact for those who experienced dry shocks. This is consistent with wealth and fertility patterns observed elsewhere.[11] As incomes rise, people tend to have fewer children for a variety of reasons, ranging from better access to contraception and family planning information to a reduced risk of infant mortality. Experiencing a large wet shock in infancy also increases the interval between births—an outcome that is desirable as it means lower overall fertility rates and positive impacts on mortality and morbidity rates for both mothers and their children.[12]

Another dimension is suggestive evidence of a relationship between large precipitation shocks in infancy and adult women's acceptance of intimate partner violence (IPV). The surveys indicate that women who were born around the time of a large dry shock are more likely as adults to believe that a husband is sometimes justified in behaving violently toward his wife. The reasons for this are unclear and could reflect the association between IPV, poverty levels, and education.

What Can Be Done to Prevent Rainfall Shocks from Becoming Destiny?

The findings discussed in this chapter highlight the long-term and intergenerational consequences of rainfall shocks. Rural African women who are born around the time of abnormally low rainfall are significantly less wealthy as adults. They receive less education, are shorter, and have lower levels of empowerment. And these impacts are transmitted across generations to their offspring.

While these results are a grim outcome for some of the world's poorest people, well-designed policies may help prevent rainfall shocks from determining their destiny. Putting in place safety nets can significantly reduce the negative impacts of, for example, droughts. An example of one such program is the Productive Safety Net Programme in Ethiopia.[13] Safety nets are most likely to be successful if they are automatically triggered when shocks occur or, in certain circumstances, when they are anticipated (Clarke and Dercon 2016).

The impact of early-life rainfall shocks on a woman's height and various anthropometric failures of her children is particularly striking. While the evidence on the impact of nutritional interventions on physical growth remains equivocal, some studies have found that programs targeting improved nutrition for younger and more vulnerable populations can reduce the likelihood of stunting.[14] In particular, programs should target a child's "first 1,000 days"—from the start of pregnancy until age two.

Clearly, there are significant gender-based dimensions to the current findings. Increasing women's empowerment through, for example, education and access to finance may help to counteract these impacts, and may have positive effects on their children, dulling the intergenerational transmission of shocks.

Resilience to rainfall shocks could also be built by introducing alternative, more drought-resistant crops, and by improved livestock management that decreases vulnerability to rainfall shortages. Increased financial inclusion and access to credit can give families the ability to save and borrow money, which provide resilience in situations where the rains do not come and crops fail. Crop and livestock insurance can also provide safety nets against unpredictable rainfall, while also boosting investment in improved agricultural practices. Indeed, studies have shown that in Ghana, providing farmers with access to rain-indexed insurance boosts investment in agricultural practices (Karlan et al. 2014).

Ultimately, economic development is the best form of resilience. But growth and prosperity are solutions for the long term.[15] The most vulnerable members of society cannot wait, particularly those living in the most hostile and drought-prone areas. In the short run, a fail-safe approach to prevent the extreme, long-term impacts of rainfall shocks will require rolling out and scaling up safety net programs. There is a growing recognition that these can constitute a vector of interventions to increase household resilience to climate variability—as demonstrated, for example, by the Sahel Adaptive Safety Net Program.

The current chapter sheds new light on just how drastic and wide-ranging the implications of neglect can be. In providing evidence for the long-term costs and risks of permanent damage, the analysis also suggests that measures taken to prevent these impacts from crystallizing as intergenerational destinies are likely more important than is often presumed.

Notes

1. As shown by Alderman, Hoddinott, and Kinsey (2006), Maccini and Yang (2009) and Carrillo, Fishman, and Russ (2015), among others. Seminal work on this topic was carried out by David Barker. See, for example, Barker (1990, 1995).

2. Almond and Currie (2011) and Almond, Currie, and Duque (2017) provide reviews.

3. For example, Almond and Mazumder (2011) find that fasting during pregnancy in observance of Ramadan negatively impacts future outcomes of the child in utero, with particularly large impacts on cognitive ability.

4. In a study of the long-term impacts of rainfall in Indonesia, Maccini and Yang (2009) find a significant impact for women who experienced above-average rainfall in infancy, but not for men. Other evidence that male and female children do not feel the long-term impacts of environmental conditions equally is provided by Carrillo, Fishman, and Russ (2015).

5. http://www.dhsprogram.com/.

6. The average woman in the sample has a mere 3.5 years of schooling, and experiencing a large negative rainfall shock in infancy results in 0.2 fewer years of formal education.

7. http://databank.worldbank.org/data/reports.aspx?source=world-development-indicators.

8. It is worth noting, however, that waterborne health impacts do not seem likely, as diarrhea, for example, is usually worse during floods and the analysis does not suggest any negative effects of significantly above-average rainfall. On the other hand, ill health could be the result of nutritional deprivation or other stresses.

9. As measured using the Composite Index of Anthropometric Failure outlined by Nandy and Svedberg (2012).

10. Recent research includes the work of Tan, Zhibo, and Zhang (2014) and Caruso and Miller (2015).

11. On the one hand, couples may respond to adverse conditions by limiting births, as discussed by Lindstrom and Berhanu (1999). On the other hand, because of the "risk insurance" demand for children, adverse economic conditions may increase that demand for children (Bongaarts and Cain 1981).

12. The relationship between birth spacing and maternal and infant outcomes has been discussed by many authors. See, for example, Omran and Standley (1976), de Sweemer (1984), and Razzaque et al. (2005). An overview of the benefits of increased birth spacing for mothers and infants is provided by Smith et al. (2009), while an overview of the causes and consequences of high fertility is presented in Casterline (2010).

13. As noted by Hobson and Campbell (2012). The Productive Safety Net Programme in Ethiopia aims to increase the resilience of poor households in rural areas in the face of food shortages.

14. A discussion of the economics of stunting is provided by Galasso and Wagstaff (2017).

15. As noted by Clarke and Dercon (2016).

References

Alderman, H., J. Hoddinott, and B. Kinsey. 2006. "Long Term Consequences of Early Childhood Malnutrition." *Oxford Economic Papers* 58 (3): 450–74.

Almond, D., and J. Currie. 2011. "Killing Me Softly: The Fetal Origins Hypothesis." *Journal of Economic Perspectives* 25 (3): 153–72.

Almond, D., and B. A. Mazumder. 2011. "Health Capital and the Prenatal Environment: The Effect of Ramadan Observance during Pregnancy." *American Economic Journal: Applied Economics* 3 (4): 56–85.

Almond, D., J. Currie, and V. Duque. 2017. "Childhood Circumstances and Adult Outcomes: Act II." NBER Working Paper No. w23017. National Bureau of Economic Research, Cambridge, MA.

Barker, D. J. 1990. "The Fetal and Infant Origins of Adult Disease." *British Medical Journal* 301 (6761): 1111.

———. 1995. "Fetal Origins of Coronary Heart Disease." *British Medical Journal* 311 (6998): 171.

Bongaarts, J., and M. Cain. 1981. "Demographic Responses to Famine." Center for Policy Studies Working Paper No. 77, Population Council, New York.

Carrillo, P., R. Fishman, and J. D. Russ. 2015. "Long-Term Impacts of High Temperatures on Economic Productivity." Working Paper No. 2015-18, Institute for International Economic Policy, George Washington University, Washington, DC.

Caruso, G., and S. Miller. 2015. "Long Run Effects and Intergenerational Transmission of Natural Disasters: A Case Study on the 1970 Ancash Earthquake." *Journal of Development Economics* 117: 134–50.

Case, A., and C. Paxson. 2008. "Stature and Status: Height, Ability, and Labor Market Outcomes." *Journal of Political Economy* 116 (3): 499–532.

Casterline, J. 2010. *Determinants and Consequences of High Fertility: A Synopsis of the Evidence—Portfolio Review*. Washington, DC: World Bank.

Clarke, D. J., and S. Dercon. 2016. *Dull Disasters? How Planning Ahead Will Make a Difference*. Oxford: Oxford University Press.

de Sweemer, C. 1984. "The Influence of Child Spacing on Child Survival." *Population Studies* 38 (1): 47–72.

Galasso, E., and A. Wagstaff. 2017. "The Economic Costs of Stunting and How to Reduce Them." World Bank Policy Research Note, World Bank, Washington, DC.

Gowin, E. B. 1915. *The Executive and His Control of Men*. New York: Macmillan.

Hallegatte, S., M. Fay, M. Bangalore, T. Kane, and L. Bonzanigo. 2015. *Shock Waves: Managing the Impacts of Climate Change on Poverty*. Washington, DC: World Bank.

Hobson, M., and L. Campbell. 2012. "How Ethiopia's Productive Safety Net Programme (PSNP) Is Responding to the Current Humanitarian Crisis in the Horn." *Humanitarian Exchange Magazine* 53: 8–11.

Institute of Medicine. 2001. "Reproductive and Developmental Effects." In *Clearing the Smoke: Assessing the Science Base for Tobacco Harm Reduction*, edited by K. Stratton, P. Shetty, R. Wallace, and S. Bondurant, 172–74. Washington, DC: National Academies Press.

Karlan, D., R. D. Osei, I. Osei-Akoto, and C. Udry. 2014. "Agricultural Decisions after Relaxing Credit and Risk Constraints." *Quarterly Journal of Economics* 129 (2): 597–652.

Kocornik-Mina, A., T. K. McDermott, G. Michaels, and F. Rauch. 2015. "Flooded Cities." CEP Discussion Paper No. 1998, Centre for Economic Performance, London School of Economics.

Lindstrom, D. P., and B. Berhanu. 1999. "The Impact of War, Famine, and Economic Decline on Marital Fertility in Ethiopia." *Demography* 36 (2): 247–61.

Maccini, S. L., and D. Yang. 2009. "Under the Weather: Health, Schooling, and Economic Consequences of Early-Life Rainfall." *American Economic Review* 99 (3): 1006–26.

Martorell, R. 1999. "The Nature of Child Malnutrition and Its Long-Term Implications." *Food and Nutrition Bulletin* 20 (3): 288–92.

Nandy, S., and P. Svedberg. 2012. "The Composite Index of Anthropometric Failure (CIAF): An Alternative Indicator for Malnutrition in Young Children." In *Handbook of Anthropometry*, edited by Victor Preedy, 127–37. New York: Springer.

NCD Risk Factor Collaboration. 2016. "A Century of Trends in Adult Human Height." *Elife* 5: e13410.

Omran, A. R., and C. C. Standley. 1976. *Family Formation Patterns and Health*. Geneva: World Health Organization.

Prendergast, A. J., and J. H. Humphrey. 2014. "The Stunting Syndrome in Developing Countries." *Paediatrics and International Child Health* 34 (4): 250–65.

Razzaque, A., J. Da Vanzo, M. Rahman, K. Gausia, L. Hale, M. A. Khan, and A. H. M. G. Mustafa. 2005. "Pregnancy Spacing and Maternal Morbidity in Matlab, Bangladesh." *International Journal of Gynecology & Obstetrics* 89 (Supp1): S41-9.

Resnik, R. 2002. "Intrauterine Growth Restriction." *Obstetrics & Gynecology* 99 (3): 490–96.

Silventoinen, K. 2003. "Determinants of Variation in Adult Body Height." *Journal of Biosocial Science* 35 (2): 263–85.

Smith, R., L. Ashford, J. Gribble, and D. Clifton. 2009. *Family Planning Saves Lives, Fourth Edition.* Washington, DC: Population Reference Bureau.

Sundet, J. M., K. Tambs, J. R. Harris, P. Magnus, and T. M. Torjussen. 2005. "Resolving the Genetic and Environmental Sources of the Correlation between Height and Intelligence: A Study of Nearly 2600 Norwegian Male Twin Pairs." *Twin Research and Human Genetics* 8 (4): 307–11.

Tan, C. M., T. Zhibo, and X. Zhang. 2014. "Sins of the Fathers: The Intergenerational Legacy of the 1959–1961 Great Chinese Famine on Children's Cognitive Development." Discussion Paper 01351, International Food Policy Research Institute, Washington, DC.

UNEP. 2010. "Africa Water Atlas." Division of Early Warning and Assessment (DEWA). United Nations Environment Programme (UNEP). Nairobi, Kenya.

World Bank. 2011. *World Development Report 2012: Gender Equality and Development.* Washington, DC: World Bank.

———. 2015. *World Development Indicators 2015.* World Bank, Washington, DC.

4

Water in the City: Drops, Blocks, and Shocks

You take delight not in a city's seven or seventy wonders, but in the answer
it gives to a question of yours.

— Italo Calvino, *Invisible Cities*

Key Chapter Findings

- Cities are vulnerable to the impacts of water shocks, and inadequate water
  services are often at the heart of the problem.

- In the cities of Latin America, moderate rainfall shocks have no discernible
  economic impacts, but large ones do—particularly dry ones. These tend to
  affect vulnerable workers, causing an estimated income loss of US$40 per
  worker per month, equivalent to about 10 percent of mean monthly income.

- Rainfall shocks can be transmitted through multiple pathways—by
  impeding firm performance and reducing sales, as well as through health
  impacts and power outages.

- Informal firms suffer from multiple risks; they are more exposed to water
  shortages and are also more vulnerable to their impacts.

- Key to addressing these challenges is improving the services of water
  utilities, which crucially depends upon better regulation.

A majority of the world's population currently resides in cities. By 2050
over two-thirds of the world will live in urban areas, and much of this
transition will occur in developing-country cities with populations of at
least 1 million (Christiaensen and Kanbur 2016; UN 2015). The urban
landscape is undergoing unprecedented shifts, with rising populations and

accelerating economic activity. Currently, cities, their firms, and their workers generate more than 80 percent of global GDP, and the continuing march of urbanization has important implications for economic growth, the reduction of poverty, and sustainability.

Cities are thirsty places. With a high concentration of assets and people, damage and disruption to economic activity in a city could have wide-ranging impacts that are transmitted through the economy. Water stress is emerging as a growing and at times underappreciated challenge in many countries of the developed and developing worlds. One in four cities, with a total of US$4.2 trillion in economic activity, is classified as water-stressed (Christiaensen and Kanbur 2016; UN 2015). Moreover, 150 million people live in cities with perennial water shortages, defined as having less than 100 liters per person per day of sustainable surface water or groundwater. In coming years, population growth and continuing urbanization will bring a 50–70 percent rise in the demand for water in cities (2030 Water Resources Group 2009). This will be fueled not only by the growing numbers of urban dwellers but also by lifestyles and consumption patterns that are more water-intensive. By 2050, almost 1 billion urban dwellers will live in water-stressed cities. With climate change, water shortages will proliferate to other parts of the world, potentially affecting even more city dwellers.

Though water scarcity is a growing problem in cities, it is floods that have captured the attention of policy makers and researchers. Floods are visible, high-impact events that destroy infrastructure, damage homes, and disrupt livelihoods. They provide compelling media material and are difficult to ignore. A flood typically triggers a rapid policy response through the panoply of relief systems available within countries, with the backstop support of international relief agencies. Perhaps as a consequence, recent research suggests that even when the short-term impacts of a flood are severe and alarming, economic recovery can come rapidly too.[1] Droughts, on the other hand, are spatially diffuse, harder to identify, and more complicated to comprehend. Their damage in an urban context emerges gradually and less visibly. Research suggests that it takes 3.5 more deaths in a drought to get the same level of media coverage as a flood.[2] As a result, less is known about the effects of droughts on city dwellers and the urban economy.

With intensifying water scarcity across cities, understanding the impacts of droughts in urban areas, and developing ways to mitigate those effects, remains an urgent task. The drought that affected São Paulo in 2014–15 was striking and revealing, both for the preventable nature and size of its impacts, and the economic importance of that city to Brazil's economy.

Figure 4.1 illustrates some of the pathways through which water deficits or excess might affect cities and their residents. This chapter explores these channels and investigates their impacts on city residents and urban firms. Since the vulnerability of a city to water-related risks is most often a consequence of either an infrastructure deficit or regulatory

FIGURE 4.1 **Possible Pathways of Exposure to Water-Related Impacts**

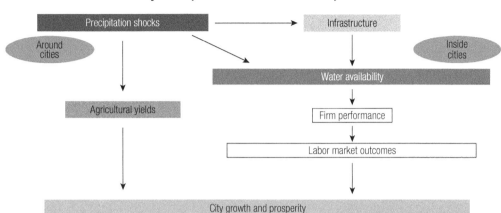

deficiencies, or some combination of these, this chapter also investigates the challenges of managing water-related infrastructure and delivering adequate water supply and sanitation (WSS) services to city dwellers.

The first part of the analysis is based on Latin America, which is the most urbanized region in the developing world and provides useful lessons to other regions at earlier stages in the urbanization cycle.[3] Data on monthly labor income from labor force surveys were used to measure the impact of rainfall shocks on incomes in the urban sector. Labor market indicators are closely tied to economic activity in a city, and consequently provide a useful metric to capture some of the broader effects of rainfall fluctuations on the urban economy.[4] The data set covers 13 million individuals living in 78 major metropolitan areas across 10 countries (Argentina, Brazil, Chile, Colombia, El Salvador, Ecuador, Mexico, Paraguay, Peru, and Uruguay). Map 4.1 shows the cities included in the empirical analysis, overlaid with the incidence of rainfall shocks over the period of investigation, from 1990 to 2013.[5] The sample covers a variety of climatic zones, with cities that have encountered both wet and dry shocks.

Water Shocks in the Cities: Drowned or Dehydrated?

The evidence suggests that the overall impact of wet shocks in cities on residents' incomes—even severe ones—is relatively modest. Moderate wet shocks—rains that are somewhat (1 standard deviation) heavier than usual (the long-run mean)—have little or no detectable impact on economic activity. On the other hand, large wet shocks (those that are 2 standard deviations greater than the long-run mean, and therefore have a greater probability to have caused flooding) lead to a small decline in monthly incomes of about 4 to 5 percent and annual labor income of about 1.5 percent, affecting formal and informal sector workers alike.

MAP 4.1 **Wet and Dry Shocks in Latin America and the Caribbean, 1990–2013**

Note: LABLAC = Labor Force Survey Database for Latin America and the Caribbean; SEDLAC = Socio-Economic Database for Latin America and the Caribbean.

Turning next to dry shocks, the analysis finds that moderate dry shocks too have no discernible impact on labor market outcomes. However, large and protracted dry shocks have significant effects that disproportionately affect informal workers, who tend to be among the poorest and most vulnerable. While formal sector workers lose about 7 percent of their labor income during large dry shocks; informal workers, the self-employed, and workers in small firms suffer a decline in labor income of about 8 to 11 percent of labor income. This corresponds to an average estimated loss of US$30 to US$45 per worker, per episode.

In sum, dry shocks have a much larger impact than large wet shocks on labor market outcomes. Notably, the income losses from large dry shocks are two to four times greater than those of wet shocks. This clearly suggests the need for greater policy attention to the often-underappreciated consequences of dry shocks in cities.

The Unexpected Pathways through Which Shocks Impact Incomes

Understanding the multiple pathways through which these shocks do their harm can provide valuable insights into how to prevent or mitigate their impacts. Much of the existing literature on rainfall shocks and cities has focused on the way droughts impact cities through rural-to-urban migration, especially in Africa (Henderson et al. 2017). For countries and regions where agriculture is the dominant sector of the economy, migration could be expected to be an important pathway through which water shocks reach cities; most obviously, people impoverished by a drought come to cities seeking relief. But with higher average income levels and the widespread use of (albeit imperfect) safety nets, other mechanisms could also be significant in Latin America.

Although not conclusive, the results suggest that in Latin America, there is no statistically discernible evidence that either wet or dry deviations of rainfall in rural areas generate any impact in adjoining cities. The reasons are unclear and warrant further investigation. But this finding could be a consequence of the protection offered by safety nets in rural Latin America. The result is also consistent with the relatively higher income levels in the region, which enable households to absorb these shocks without migrating to cities.[6]

Instead, rainfall shocks bring consequences that are city-specific. One pathway identified is through health; large dry shocks coincide with a rise in the total number of hospital admissions, and the incidence of diarrheal diseases in children under two.[7] This is consistent with the epidemiological literature, which links droughts to health impacts caused by a decline in water quality. The range of diseases associated with droughts includes cholera, typhoid, hepatitis, and leptospirosis, among others (Stanke et al. 2013). A second channel is through the level of economic activity. Droughts are associated with a higher incidence of power outages, which would dampen overall activity in the affected cities.[8] These and other factors could have impacts on the performance of firms in cities, which is the issue that is analyzed next.

Water in the Business Environment

Firms are a critical engine of city growth. They generate jobs, provide essential products and services, encourage innovation, and link cities to global markets. Cities have always been centers of commerce and trade, providing sizable markets for businesses and access to a pool of resources—both human and physical capital. This symbiotic relationship is expected to deepen even further. The second half of the 20th century saw a trend of decentralization as businesses moved away from cities to the suburbs. Over the past two decades, especially in developed economies, businesses have reversed that trend, moving back to the city to engage younger talent seeking better standards of living.[9] Infrastructure is central to this symbiotic relationship between cities and firms, and it determines the physical form of a city—with consequences for productivity, incomes, livability, social inclusion, and resilience.

Water is a critical input in production and is often taken for granted. Hence, for instance, access to water and water infrastructure does not feature in the World Bank's Ease of Doing Business index. Nor is there much attention given to the role of water in city development in the academic literature. This section explores whether the availability of water has an impact on firm performance, using data from nationally representative enterprise surveys data for over 16,000 formal firms across 103 economies between 2009 and 2015. The surveys provide information on the reliability of water and the duration of water outages.

Water shortages experienced by firms exhibit a marked pattern across all regions. Smaller firms and those located in low- or middle-income countries are most exposed to water shortages (figures 4.2 and 4.3). On average, firms in low- and lower-middle-income economies face more than twice the number shortages than those in upper-middle and high-income economies (nine versus four). Incidents of water shortages also tend to last longer on average in low- and lower-middle-income economies. A larger fraction of small firms are affected by water outages relative to medium-sized firms (20–99 employees) or large firms (100-plus employees).

Further patterns emerge from the data. Frequent power outages are also positively correlated with a higher frequency of water outages; and unsurprisingly, large dry shocks lead to more frequent water shortages. Finally, in countries where water outages are frequent, firms are found to pay bribes to gain access to water (see box 4.1).

The Costs of Service Deficits

Water outages are found to have an impact on firm performance. An additional water outage in a typical month results in an average loss of 8.7 percent of sales. Stated differently, this implies that a 1 standard deviation increase in water outages is associated with a 1.7 percent loss of sales, or about US$200,000.

The consequences for informal firms—a ubiquitous form of economic enterprise in developing countries—are much worse. Not only do they receive poorer services with more frequent outages, but they also have less capacity to cope with service disruptions. Across 18 economies in Africa, South Asia, and Latin America, an additional water outage is associated with a 34.8 percent loss in sales. Alternatively, a 1 standard deviation increase in the number of water outages per day in the last month leads to a 4.9 percent loss of sales.[10] These results are summarized in figure 4.4.

FIGURE 4.2 **Water Shortages and Development (Manufacturing Firms), 2009–15**

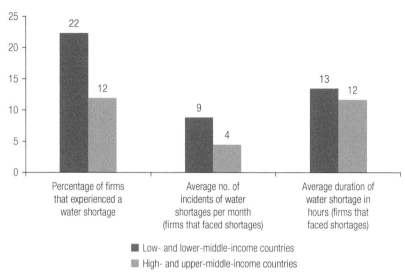

Source: Calculations based on Enterprise Surveys.

FIGURE 4.3 **Water Shortages and Firm Size (Manufacturing Firms), 2009–15**

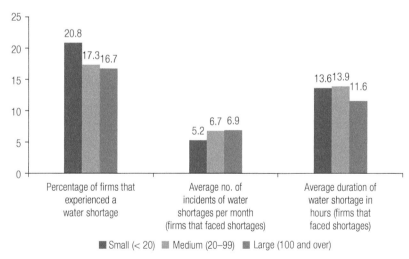

Source: Calculations based on Enterprise Surveys.

BOX 4.1. Governance and Water Service Delivery: A Double Jeopardy on Firms?

The provision of water is one of the key public services of a municipality, but also one of the most hobbled by mismanagement and exploitation because of the de facto monopoly power wielded by water utilities. There is also the issue of governance and how it influences water service delivery. Poor governance creates an environment in which bribery can thrive. The Enterprise Surveys data show that firms that make an informal payment or gift to obtain a water connection are more likely to face water shortages than firms that do not. Estimates indicate that 26 percent of firms experiencing water shortages made informal payments to obtain a connection, whereas only 17 percent of firms that did not experience shortages made such payments (figure B4.1.1).

There are two potential explanations for this: One is that this is a mere correlation—a poorly managed water utility is likely to deliver poor services and may also be susceptible to bribery. The other is that poor governance leads to poor water service delivery. Firms are then compelled to pay bribes to gain access to piped water services. This leads to a low-level equilibrium, in which public infrastructure is used to divert rents for private gain, rather than for improvements in, or maintenance of, infrastructure. Similar results have been found for electricity (Pless and Fell 2017).

FIGURE B4.1.1 Percentage of Firms That Paid an Informal Payment for a Water Connection, 2009–15

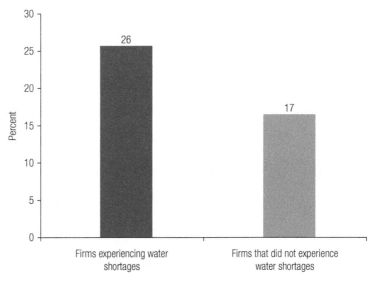

Source: Calculations based on Enterprise Surveys.

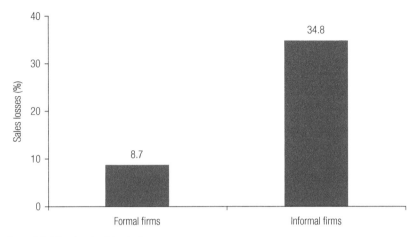

Source: Calculations based on Enterprise Surveys.

Addressing Water Shortages and Shocks: The Regulatory Lever

Water shortages are often a consequence of public policy failures compounded by market failures. When city water supply and sanitation (WSS) services struggle to cope under "normal" conditions, the pressures of rainfall shocks can magnify these stresses, leading to widespread failure. Investing in and correctly managing infrastructure are therefore key to building resilient systems that can insulate cities from the growing scarcity and variability of water availability. This section tackles another and perhaps vital dimension of the problem—that of improving the efficiency of water utilities through better-targeted incentives and regulations. There are three overarching priorities to improving service delivery: designing contracts with water utilities (whether public or private) that are *incentive compatible*; increasing competition during entry into the market; and improving access to credit and finance for infrastructure investments.

A Natural Monopoly in an Ocean of Complexity

As a natural monopoly charged with providing a "merit good" that is a declared human right, WSS utilities are seldom far from controversy. Natural monopoly problems loom large in the WSS industry. Water distribution and sewerage services are characterized by significant capital costs, and relatively high barriers to entry because of the capital-intensive nature of operations. Capital costs are high enough to make duplication of infrastructure economically unviable in virtually all circumstances, rendering the service a natural monopoly.

Inescapable advantages accrue to a natural monopolist, whether public or private, in its interaction with regulators and customers. The task of policy and regulation is to recognize these asymmetries in bargaining and to ensure that affordable, quality services go hand-in-hand with a fair and "normal" rate of return to the service provider. Failure to achieve this balance renders the city vulnerable to a host of problems, including unsustainable utilities, corruption, and poor service provision.

Contractual Design Can Improve Service Delivery under Ever-Riskier Circumstances

Critically, a service provider knows more about its own cost structure and level of efficiency than does its regulator. This informational asymmetry translates into a bargaining advantage that can lead to inadequate services, inflated costs, or the ad hoc renegotiation of contracts.[11]

By inflating costs (through inefficiency, cost padding, or simply by not revealing the true costs), a firm can capture a greater proportion of available rents. If the service provider is public, these rents could accrue in the form of higher wages and salaries, politically connected appointments, poor service, and pervasive underperformance. In the private sector, inflating costs is more likely aimed at ensuring that the utility earns an adequate rate of return.[12]

A key insight of the mechanism design literature is that with appropriate attention to contractual design, many of these problems can be at least partially ameliorated. If the main problem is that regulators do not have the necessary information (perhaps because accounting practices are poor, or because the regulator lacks capacity), it is reasonable to pick a regulatory regime that limits the need for the unavailable information. This can be achieved by setting a maximum price (termed a *price cap*) for water, informed by the cost information that might be available and by international benchmarking.[13] The utility is left free to make whatever rate of return it can within this framework. A price cap provides an incentive for the operator to minimize costs in order to maximize profits. And by doing so, it could provide the regulator with more information about the true (and hidden) costs for the next cycle of negotiation.

But when the main concern is the regulator's lack of accountability, due either to political interference or corruption, price caps would be highly inappropriate and could become a tool for manipulation and rent extraction by self-serving bureaucrats. In this case, a more appropriate contract would limit the allowable rate of return by defining a maximum markup over audited costs (termed *cost-plus*), complemented with international cost benchmarking (Auriol and Blanc 2009).

When both concerns apply equally, offering a menu of choices may be the best solution. In this case, a low-cost firm would prefer a price cap, and a provider with higher costs or more uncertainty about future costs would rather have the cost-plus option. In expressing a preference, firms reveal information about their cost structures and comparative advantages, which allows for better-informed regulation (Guasch et al. 2016).

Competition for the Market Is Especially Important

Procurement rules matter, too. Since market forces play a very limited role in promoting service quality and cost efficiency in a natural monopoly, competitive pressures can be brought to bear when firms bid for the right to provide water service to a city. This is competition *for the market*, since there can be no significant competition *within the market*. If successful, the process can deliver many of the benefits of a competitive market. Greater cost efficiency would be achieved because the firm that is able to pay the most for the license is also the firm with the lowest cost. At the same time, monopoly rents could be distributed to consumers, not the firm, if service quality and levels are adequately defined in the bidding process.

In practice, competition for the market has been circumscribed by the dominance of a small number of large players in the sector that specialize in different regions. For instance, evidence suggests that smaller municipalities in France (communities with fewer than 10,000 inhabitants) pay significant price premiums for water, even after accounting for scale economies (Chong, Saussier, and Silverman 2015). In addition, the type of procurement process that is used can determine the number of bidders and the price of bids. Complex proposals, where treatment plants and a network component are bundled, have been found to elicit fewer bids at a greater cost than simpler, unbundled bids (Estache and Iimi 2011).

Overall, enhancing competition for the market calls for processes that can increase the number of potential bidders. Where feasible, this might imply increasing the scope for new or smaller local actors to enter the market; a simpler unbundling of contracts could be another part of the solution.

A further insight from the literature is that, even with skewed information, regulation and contracting can be improved by targeting the specific constraints that dominate a particular market.[14] Procurement rules can be designed to reveal greater information about costs. And since regulators are stymied by a lack of access to accounting data, obligations to disclose information could improve accountability and service outcomes.

Conclusion

Cities tend to be better insulated from the effects of water scarcity than rural communities. Their infrastructure allows them to draw on geographically remote sources, which buffers them against the effects of overextraction and droughts. Policy makers also often prioritize cities' water needs over those of rural communities, notably because of the economic value of urban centers. For this reason, moderate rainfall shocks (wet or dry) tend not to affect cities, unlike rural areas. It takes prolonged shocks before cities suffer from rainfall-induced water shortages, by which time the costs are high. At that point, short-term remedial options are limited, because dwindling water resources are strained by multiple and competing users. Smart regulation of water utilities can play a vital role in ensuring that cities get the water they need, in both quantity and quality. But attention must also be paid to the wider context of the problem, both spatial and climatic, through watershed protection and integrated water management that better connects rural and urban areas.

Notes

1. One exception is a recent unpublished paper by Acevedo (2015) on urban Colombia. The vulnerability of cities to floods is well documented, and there is an extensive literature on disaster risk assessment. This vulnerability is particularly high in coastal cities. Some estimates put the cost of average global flood losses at approximately US$6 billion per year, increasing to US$52 billion by 2050 based on projected socioeconomic change alone (Hallegatte et al. 2013). Factoring in climate change and subsidence, the estimates in this chapter show that current protections would need to be upgraded to address losses of US$1 trillion or more per year. There is also a growing body of empirical literature that finds that recovery from a flood is typically rapid. Kocornik-Mina et al. (2015), looking at 1,800 cities between 2003 and 2008, found that large-scale floods (those displacing more than 100,000 people) reduced nighttime lights, a proxy for economic activity, by 2–8 percent within cities in the year of the flood.

2. http://perseus.iies.su.se/~dstro/Disasters.pdf.

3. Latin America is the second-most urbanized region in the world, after North America (82 versus 80 percent), and ranks ahead of Europe (UN ESA 2014).

4. De la Torre et al. (2015, 2016) and Calvo-Gonzalez et al. (2017) note that during the boom period of 2003–08, growth in labor income translated into more than 2 percentage points of poverty reduction per year. Growth in all other income sources, including public transfers, accounted for less than 1 percentage point per year. During the slowdown of 2012–14, continued yet slowing growth in labor income accounted for practically all poverty reduction. This means that while labor income growth declined, its importance for poverty reduction in fact grew. That such an important input to poverty reduction and shared prosperity could be threatened by water shocks, even in a region with comparatively good infrastructure, highlights the urgency of the issue.

5. The definition of shocks follows that defined in chapter 2 and described in box 2.1. The analysis is based on monthly and annual micro data from the Labor Force Survey Database for Latin America and the Caribbean (LABLAC; lablac.econo. unlp.edu.ar/) and from the Socio-Economic Database for Latin America and the Caribbean (SEDLAC; sedlac.econo.unlp.edu.ar/), a household-level data set. These are combined with rainfall variables aimed at capturing climate-related water shocks on the labor outcomes of workers of major cities across Latin America. The sample is representative of about 300 million households living in 78 of the largest metropolitan areas of nine Latin American countries (Brazil, Chile, Colombia, El Salvador, Ecuador, Mexico, Paraguay, Peru, and Uruguay). It covers a period of more than 20 years (1992–2013) from the annual data included in SEDLAC and eight years (2005–2013) from the monthly data of LABLAC.

6. This is established through a simultaneous equation model that studies the impact of rainfall shocks on net primary productivity (NPP) around cities and the impact of shocks and NPP changes on incomes in cities. The results of the model suggest that the important negative impact of droughts found in cities does not seem to be driven by changes in the agricultural sector but by dynamics that are specific to cities themselves. Further details of the methodology used can be found in volume 2 of this book (*Shocks in the Cities* chapter).

7. This analysis shows that water shocks, both moderate dry and wet shocks as well as large dry shocks, increase the number of hospital admissions. The effect of

contemporary moderate shocks remains limited (0.5 percent increase), while large dry shocks result in a 5 percent increase in hospital admissions in Brazil. The investigation is further deepened to look at the occurrence of diarrhea in children under five years of age. Diarrheal diseases of children under five are a good proxy, as they can be expected to reflect a constraint on household members who may themselves be affected or need to care for sick infants. While the data reflect more extreme cases requiring hospitalizations, a spike in this type of case following a shock could be expected to also be associated with a surge of less severe diarrheal diseases that may not require hospitalization and yet be disruptive to households' activities. Results show an increase of cases of diarrhea regardless of the type of shocks (wet or dry), with a marked increase in the case of large dry shocks (6 percent versus 2 percent, in the case of large wet shocks).

8. When controlling for firms' characteristics, the findings also suggest that moderate shocks increase the number of power outages. But this increase in outages is 10 times lower than for large dry shocks, indicating that the transmission channel of rainfall shocks through electricity to the labor market is predominantly due to large dry shocks.

9. See, for example, Clark and Moir (2014), Schwartz (2016), and Wieckowski (2010).

10. Some caution should be employed in interpreting results. For one, the surveys are not representative of the informal economy. Second, questions on water outages were asked of firms that used water for business activities. Despite data deficits, this is an entirely new area of research that merits further study.

11. Contract renegotiations are deemed to be a symptom of design and regulatory deficits. Recent empirical work suggests that this is not uncommon in the WSS sector. For instance, empirical assessments suggest that in LAC between 1990 and 2010, more than 90 percent of public-private partnership (PPP) arrangements that were contracted using standardized regulatory templates were renegotiated. The reasons vary and include changing political priorities, but happen most often because of a failure to recognize the country-specific constraints and the incentives imparted by informational asymmetries (Guasch et al. 2016). An implication is that the standardization of advice that neglects country-specific characteristics and how these interact with the informational advantages of a service provider can be counterproductive.

12. Bank of America/Merrill Lynch (2014) reports that investors have recognized that the WSS sector offers a higher and less-variable inflation-protected return than other infrastructure investments. Moreover, these investments are decoupled from economic growth, and act as an alternative to low bond yields and volatile equity markets. The S&P Global Water index (SPGYAQD)—covering 50 global companies involved in water-related businesses—has consistently outperformed a range of sample benchmark indices with a five-year annualized return of close to 20 percent (see Bank of America/Merrill Lynch, 2014, Blue Revolution—Global water primer; http://www.merrilledge.com/publish/content/application/pdf /gwmol/Themative-Investing-Global-Water-Primer.pdf).

13. In this discussion, the service level and quality are held constant. The argument extends to these dimensions, too.

14. This is of special importance when dealing with riskier situations, such as the expansion of access to more vulnerable groups or to areas of greater exposure to climatic shocks.

References

2030 Water Resources Group. 2009. "Charting Our Water Future." http://www.mckinsey. com/business-functions/sustainability-and-resource-productivity/our-insights /charting-our-water-future.

Auriol, E., and A. Blanc. 2009. "Capture and Corruption in Public Utilities: The Cases of Water and Electricity in Sub-Saharan Africa." *Utilities Policy* 17 (2): 203–16.

Bank of America/Merrill Lynch. 2014. "Blue Revolution—Global Water Primer." http://www.merrilledge.com/publish/content/application/pdf/gwmol/Thematic-Investing-Global-Water-Primer.pdf.

Calvo-González, O., R. A. Castaneda, M. G. Farfán, G. J. Reyes, and L. D. Sousa. 2017. "How Is the Slowdown Affecting Households in Latin America and the Caribbean?" World Bank Policy Research Paper No. 7948, World Bank, Washington, DC.

Chong, E., S. Saussier, and B. Silverman. 2015. "Water under the Bridge: Determinants of Franchise Renewal in Water Provision." *Journal of Law, Economics and Organization* 31 (1): 3–39.

Christiaensen, L., and R. Kanbur. 2016. "Secondary Towns and Poverty Reduction: Refocusing the Urbanization Agenda." Policy Research Working Paper, No. 7895.

Clark, G., and E. Moir. 2014. "The Business of Cities." *Future of Cities*, blog post, October 17. Government Office for Science, London. https://futureofcities.blog.gov.uk/2014/10/17/the-business-of-cities/.

De La Torre, A., A. Ize, and S. Pienknagura. 2015. "Latin America Treads a Narrow Path to Growth: The Slowdown and its Macroeconomic Challenges." LAC Semiannual Report, World Bank, Washington, DC.

De La Torre, A., F. Filippini, and A. Ize. 2016. "The Commodity Cycle in Latin America: Mirages and Dilemmas." LAC Semiannual Report, World Bank, Washington, DC.

Estache, A., and A. Iimi. 2011. "(Un)Bundling Infrastructure Procurement: Evidence from Water Supply and Sewage Projects." *Utilities Policy Elsevier* 19 (2): 104–14.

Guasch, J. L., D. Benitez, I. Portabales, and L. Flor. 2016. "The Renegotiation of Public Private Partnership Contracts (PPP): An Overview of Its Recent Evolution in Latin America." *Revista Chilena de Economía y Sociedad* 10 (1): 42–63.

Hallegatte, S., A. Vogt-Schilb, M. Bangalore, and J. Rozenberg. 2017. *Unbreakable: Building the Resilience of the Poor in the Face of Natural Disasters*. Washington, DC: World Bank.

Henderson, J. V., A. Storeygard, and U. Deichmann. 2017. "Has Climate Change Driven Urbanization in Africa?" *Journal of Development Economics* 124: 60–82.

Pless, J., and H. Fell. 2017. "Bribes, Bureaucracies, and Blackouts: Towards Understanding How Corruption at the Firm Level Impacts Electricity Reliability." *Resource and Energy Economics* 47: 36–55.

Schwartz, N. 2016. "Why Corporate America Is Leaving the Suburbs for the City." *New York Times*, August 1. https://www.nytimes.com/2016/08/02/business/economy/why-corporate-america-is-leaving-the-suburbs-for-the-city.html.

Stanke, C., M. Kerac, C. Prudhomme, J. Medlock, and V. Murray. 2013. "Health Effects of Drought: A Systematic Review of the Evidence." *PLoS Currents Disasters*. June 5. Edition 1.

UN (United Nations). 2015. "World Population Prospects: The 2015 Revision, Key Findings and Advance Tables." Working Paper No. ESA/P/WP.241. UN Department of Economic and Social Affairs, Population Division.

Wieckowski, A. 2010. "Back to the City." *Harvard Business Review*. May. https://hbr.org/2010/05/back-to-the-city.

5
Going with the Flow:
The Policy Challenge

Human nature is like water, it takes the shape of its container.
—Wallace Stevens, *Collected Poems*

Key Chapter Findings

- Preparing for future rainfall variability and water scarcity will require a host of responses to build buffers and reduce exposure to risks. This will need to include supply-side investments with new technologies, as well as demand-side policies and complementary approaches.

- Greater policy attention needs to be given to the *paradox of supply*—that when water is made available free or well below its value, it is used unsustainably, resulting in scarcity and vulnerability to shocks.

- Traditional options for expanding water supplies—including better storage infrastructure, desalinization, and water reuse—are vital for meeting future water requirements.

- Increasing supplies must be coupled with policies that serve to limit water demand, and could include water pricing and water permit trading, with appropriate safeguards to ensure affordability, access for all, and adequate flows for the environment and to assure sustainability.

- Weather insurance and safety net programs are important tools that should be expanded to ensure that rainfall shocks do not have long-lasting negative consequences.

- Many of the challenges related to regulating water monopolies need to be tackled at the contract design phase.

Drought and deluge are costly realities in many countries. The spectacle of a flood is so powerful a metaphor that nearly every civilization seems to have a deluge myth.[1] Floods continue to attract vast amounts of media, scholarly, policy, and financial attention. Yet this book demonstrates that droughts are often the more damaging event. They are misery in slow motion, with farms, firms, and families suffering larger, longer-lasting, and more insidious impacts than previously understood.

On the farm, one consequence is predictable—a drought depresses agricultural yields. Less predictably, droughts induce a considerable amount of cropland expansion, adding to pressures on forests. In the city, water scarcity and unreliable infrastructure stall production, sales, and the incomes of workers. Perhaps of greatest concern are the hidden and irreversible human costs. In rural Africa, a rainfall shock in infancy becomes a person's destiny, often resulting in cognitive and physical impairments that translate into less education, less wealth, less height, and even harmful intergenerational consequences for their offspring. To make matters worse, more than 60 percent of humanity already lives in areas of water stress, where available supplies cannot sustainably meet demand (Mekonnen and Hoekstra 2016). If water is not managed more prudently, the crises and stresses of today will become the catastrophes of tomorrow.

Confronted with intensifying levels of scarcity and shocks, policy makers need to take a long-term view, and thus resist the impulse to make decisions and investments that might solve a specific or current problem but sow the seeds for a larger crisis down the line. A prime example of this is demonstrated in chapter 2, where it was shown that investments in irrigation infrastructure in already-water-scarce areas might seem like a prudent investment decision in times of plenty, but when dry shocks inevitably occur, the infrastructure fails and livelihoods can be destroyed.

Just as a financial crisis can shake markets and ruin pensions, water shocks have the same power to slow production and destroy livelihoods. To avoid these impacts, water must be managed prudently throughout its entire economic cycle—from its source; to the infrastructure that distributes it to farms, firms, and families; and back to the lakes, rivers, and aquifers where it began (figure 5.1). At its source, water is a public good that is vulnerable to overexploitation and the tragedy of the commons. As it moves into pipes, it becomes simultaneously a private good and a merit good—one to which people have a right as a necessity for life and health. In cities, this dual challenge is compounded by the fact that the cost of building multiple water systems is prohibitive. Water must be supplied through a single network, which must have a single owner—a monopolist—that needs to be regulated to ensure adequate access to water at a price that people can afford. Last, the water passes through sewers and re-enters the ecosystem where, if untreated, it can pose major health and environmental risks. Neglecting these linkages can result in policy decisions that are at best less effective than they could be, and at worst downright harmful.

Successfully managing the effects of capricious rainfall levels and water levels through each stage of the economic lifecycle of water requires

FIGURE 5.1 **The Economic Cycle of Water**

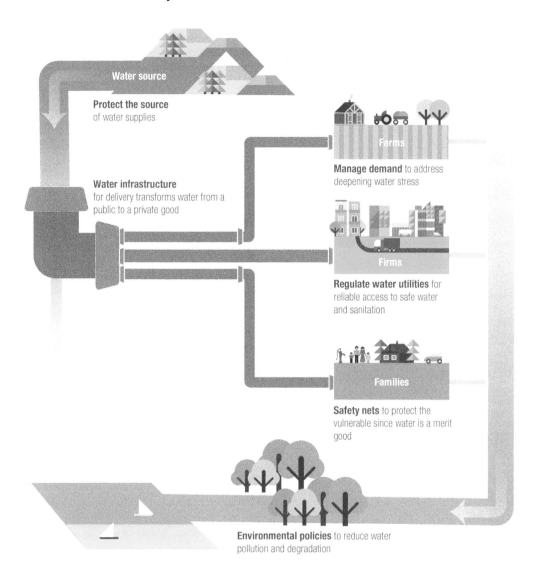

balancing three competing priorities: efficiency, equity, and sustainability. As a scarce resource, water ought to be allocated efficiently to maximize its contribution to growth, jobs, and well-being. At the same time, the allocation needs to be equitable to meet the needs of different water users and provide everyone, in both urban and rural areas, with access to clean, safe sources of water for drinking, sanitation, and other essential uses. Finally, sustainability is paramount since freshwater is a renewable natural resource upon which all life outside the oceans depends. Protecting the sources of water—the watersheds, lakes, streams, rivers, and aquifers— ensures that natural ecosystems can replenish water supplies for current and future generations and the environment. In most countries, meeting these goals will entail fundamental regulatory, legislative, and policy changes. Box 5.1 describes a conceptual framework for ensuring that water reallocation is just, allocative, and dynamically efficient (JADE). The remainder of this chapter discusses tools that policy makers should use for decreasing water-related risks at each stage in the economic cycle of water.

BOX 5.1. Just, Allocative, and Dynamically Efficient Water Management Policies

Given the need to balance multiple considerations and approaches, a just, allocative, and dynamically efficient (JADE) water allocation strategy offers policy makers a framework to guide effective water reallocation. JADE encompasses the notions of efficiency, sustainability, and equity. To meet JADE criteria, a policy would need to be allocatively efficient (spatially and between sectors), dynamically efficient (temporal), and equitable (for people and ecosystems). A water allocation would be allocatively efficient if it is not possible to reallocate water across uses to increase the sum of the net benefits (gross benefits less all direct and indirect costs). An allocation is intertemporally efficient if it is not possible to increase the net present value of all water uses (including in situ uses) through redistribution. An allocation of water is equitable if it conforms to established norms of distributive justice that encompass ecosystem uses and needs. Contrary to many definitions of misallocation that focus solely on allocative efficiency, this wider definition regards patterns of water use that are inequitable as suboptimal.

The policy implications of such an approach are clear. If there are trade-offs between efficiency and equity or sustainability, there needs to be sufficient and just compensation for the losses. The JADE approach is therefore intrinsically sustainable in recognizing that efforts to promote more efficient water allocation that fail to address externalities and equity implications are unlikely to succeed over the long term.

Achieving the JADE ideal is no simple task, particularly where weak institutional capacity and large entrenched interests persist. In these situations, reallocation and reform are likely only to unfold over long time frames, and to require iteration at multiple stages of the process, starting with simpler tasks and reforms and moving to more complex and impactful actions. Although reallocation involves navigating a set of factors that are highly dependent on context, location, and the parties involved, a three-step process can help guide policy makers. During the first phase, *Scoping*, the problem of water misallocation in a given region is assessed, including an exercise to account for water availability and stress and understand key stakeholders and interests. The second phase, *Assessment*, focuses on evaluating different water allocation responses—the gains and the losses—including the proper mix of incentives, policies, and investments. The third and final phase, *Evaluation and Adaptation*, emphasizes implementation and assessment of progress. Such assessment facilitates policy iteration in response to unanticipated challenges encountered during implementation.[a] Incremental change can slowly shift outcomes toward the ideal.

a. Grafton, Garrick, and Horne 2017.

Supply-Side Investments Are Essential

As figure 5.1 demonstrates, infrastructure is the conduit through which water is transferred from its source, where it is a public good, to its many uses and users as a private good. Little has changed over millennia in this regard. The ancient Romans built enormous networks of aqueducts to move water hundreds of miles from areas of abundance, usually high up in the mountains, to their cities. On the other side of the world, the Chola Dynasty in southern India built the Kallanai Dam in the 2nd century AD to divert water for irrigation purposes. The Kallanai Dam is still supporting local agriculture by providing irrigation benefits, and it is believed to be the world's oldest dam still in operation.

Water storage technology remains remarkably similar to the systems used in those days. Even today, the most widely used method for increasing water supply is to dam rivers. During periods of excess, dams allow for the capture of runoff, to be released during periods of deficiency. They also have the potential benefit of generating power and offering flood protection. However, as figure 5.2 shows, per capita reservoir storage has been declining since about 2000, partly because of poor management and loss of storage capacity to sedimentation (Zarfl et al. 2015). At the same time, the stage is set for a large increase in the world's number of dams, projected to rise 16 percent by 2030, with storage volume increasing by about 40 percent.[2] Estimates suggest that even an expansion of this scale may not suffice to meet future demand. In any case, the scope for increasing

FIGURE 5.2 **Per Capita Reservoir Storage Is Declining in the 21st Century**

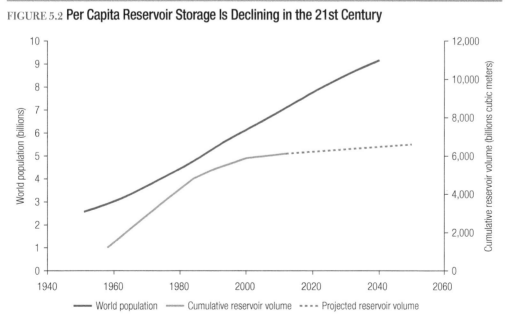

Source: Annandale et al. 2016.

storage even further is limited by the diminishing availability of good locations and an intensifying competition for scarce water.

Desalination is another way to increase water supply. It can create a virtually limitless source of drought-proof, clean water in coastal areas. However, it is highly energy-intensive and raises problems related to the disposal of brine, the by-product of desalination, and harm through the impact on fisheries and aquatic biomes. Desalination is typically not economically viable for lower-value uses, such as agriculture, but has the potential to offer back-stop protection against rainfall variability for higher-valued uses, such as manufacturing or municipal water use.

Advancing technologies offer promising opportunities for water supply expansion and water resource recovery. One such technology is wastewater recycling, which is particularly useful in large cities. But it has two major drawbacks: recycling plants tend to be energy-intensive—in the United States, they consume 2 percent of total electricity[3]—and they generate a by-product, known as sludge, which is difficult to dispose of in an environmentally benign manner. Newer technologies show promise for alleviating these problems, which would make wastewater recycling a much more viable prospect for addressing water scarcity in cities. Recent work has shown that wastewater recycling can be accomplished at net-zero energy use. Biogas, a by-product of the treatment process, is captured and used to offset the energy consumption of the facility (Chen and Chen 2013). At the same time, new sludge by-products are being developed, including fertilizer, cement, and fuel (Buchauer and Huang 2016). These advances offer exciting opportunities not just for closing the water cycle but also for reducing carbon emissions, energy costs, and environmental contaminants. Additional research will determine the commercial viability and real-world opportunities to scale up these new technologies.

An alternative but less recognized option to built infrastructure is the use of "natural capital" to bolster water supply and storage. Natural capital investments seek to enhance the ability of ecosystems and the natural environment to store more water, enhance water quality, reduce flooding, and provide other critical benefits. These benefits are frequently taken for granted, but many studies show that relatively inexpensive investments in preserving the environment and helping ecosystems function more effectively can provide high rates of return, since the investment required is often negligible. An early example is an innovative decision by New York City to acquire land in the nearby Catskills region to filter and store water in natural ecosystems, saving the city US\$6 billion in capital costs. Elsewhere, studies have shown that preserving floodplains from development can reduce flood damage by up to 78 percent; and using environmentally sensitive agricultural and land management practices can increase water flow by up to 11 percent (Watson et al. 2016; Abell et al. 2017). Importantly, natural capital investments can be complementary to infrastructure. Thus, investing in a suite of solutions—for example, protecting watersheds and forests, together with a canal or dam for irrigation—produces greater benefits than investing in any single one of these solutions (Guannel et al. 2016). While natural capital solutions may not entirely match the storage capacity of large dams, they offer the potential to address some water scarcity issues without large financial outlays or environmental damage.

Managing Water Risks in the City Requires Greater Efficiency and Investment

As figure 5.1 demonstrates, after water flows through infrastructure and into pipes heading into cities, it becomes a private good. While building the appropriate infrastructure is challenging, often a more difficult task is ensuring reliable, affordable, and safe supplies of water in cities. The task is rendered especially difficult by one of the inherent characteristics of the water industry—that it is a natural monopoly that provides an essential service, implying the need for regulation. Improving performance in the industry will call for enhancing efficiency as well as expanding service coverage. While there is no single way of achieving this, there are several issues that need to be considered in the design and contractual arrangements of water and sanitation systems.

Since there is limited scope for competition within the water market, competitive pressures can be brought during the bidding phase of competition for the market—at the procurement stage if the provider is a private firm. But even here the evidence indicates that competition is limited and most potential entrants are deterred by the size and complexity of the contracts. Simplifying procurement rules and unbundling tasks to allow for the entry of smaller firms would be useful, though it might entail a loss of scale economies.

The fundamental challenge for policy and regulation is to recognize the monopoly power of the supplier, whether public or private, and seek policies and incentive structures that target the sources of inefficiency. The mechanism design literature (briefly summarized in chapter 4) provides useful approaches that can balance the needs of consumers for adequate and affordable services with those of the service provider to earn a fair and "normal" rate of return. A key insight of this literature is that performance can be significantly improved by targeting the specific constraints that dominate a particular market through appropriately designed regulations and contracting incentives.

Closing the Supply Gap Requires Demand Management

Going forward, countries will need a significant increase in the full menu of available infrastructure facilities and services. But caution is also warranted. Supply-side interventions, while essential, will not on their own resolve water management problems. Paradoxically, infrastructure that is built to ameliorate dry shocks can create the very conditions that magnify and accentuate their adverse impacts. The reasons for this would have been familiar to 19th-century classical economists schooled in Say's Law—the proposition that "supply creates its own demand." This implies that water made available cheaply will be used freely. The supply of free or underpriced water in arid areas spurs the cultivation of water-intensive crops such as rice, sugarcane, and cotton, which in turn increases vulnerability to drought and magnifies the impacts of dry shocks. One well-known study found that access to the Ogallala aquifer in the United States induced a shift to water-intensive crops that increased drought sensitivity over time.[4] The Aral Sea is a more extreme example of a resource that has gone from abundance to depletion within a generation. To increase cotton production, the then–Soviet government diverted rivers that fed the Aral Sea, and as a result it today holds less than one-tenth of its former volume. This book demonstrates that this paradox of supply is a more widespread phenomenon than was previously known and can be found at a global scale.

Ill-considered responses are often magnified by perverse incentives created by a complex web of subsidies. Estimates suggest that global agricultural subsidies exceed US$500 million a year (Potter 2014), though the portion going to water-intensive crops is unknown. The form these subsidies take varies considerably across crops and between countries. Subsidies to sugarcane, for instance, may occur through direct price support, through support for processed outputs such as sugar and ethanol, or through subsidies to inputs such as fertilizers or the energy used to pump water. Low-yielding and water-thirsty crops in arid regions often shelter behind a variety of tariff and nontariff barriers. This partly explains the vast difference in rice yields between the dry parts of Africa and those of Southeast Asia, the rice basket of the world. Yields in Asia vary from between 4 and 6 tons per hectare, compared, for example, to 1.4 tons in the Ruaha Valley in Tanzania. Not only are yields low; the water that nurtures this rice comes at a high opportunity cost, depriving the country's capital of cheaper electricity and exacting an adverse environmental toll on the Ruaha ecosystem and its inhabitants (Damania et al. 2015).

The economically prudent strategy is to encourage production of crops that are more closely aligned with a country's natural comparative advantages, rather than to subsidize crops that enhance vulnerability to climate shocks. But change is often difficult, as the legacy of the past typically circumscribes the present. Path dependence can set in, which renders corrective actions more difficult and costly, often requiring generous compensation to enable even small changes.

Better Allocation of Water Can Increase the Size of the Pie

As illustrated in figure 5.1, demand-side interventions become more critical as water scarcity proliferates around the globe. Where water is abundant, there is little reason to restrain its use; but when demand exceeds available supply, it needs to be rationed. Rationing and reallocation can occur in three ways: through administrative decree that mandates ownership or transfer of water; through collective action in small communities or groups; or through market forces that transfer water by balancing the available supply with potential demand through pricing (Menzen and Ringler 2008). Administrative decrees may lead to inefficient outcomes when the politically connected receive a disproportionate share of water. On the other hand, collective action is typically only feasible in small communities. Therefore, interest has recently grown in market-based allocation schemes. These are fraught with challenges; but the potential payoffs, in terms of resolving scarcity challenges, can be equally large.

Despite their potentially large benefits, market-based instruments such as prices and tradable permits are often greatly feared because of anxieties about elite capture, denial of services to the poor, and complex social values regarding water. No matter which allocation regime is used, there is a need for adequate safeguards to ensure access for poor households and farmers, as well as to protect the environment itself.

The value of water today is much less than its potential, since rights, responsibilities, and liabilities are poorly defined. If a water pricing or trading regime can increase the size of the pie—through investment, innovation, and economic growth—then it should be possible to make all participants better off and in some cases even eliminate trade-offs by generating water savings. In this regard, there is very limited experience to draw upon, but evidence from Australia is encouraging. Some estimates suggest that the annual rate of return on owning a water right exceeds an astonishing 15 percent, which reflects generous government support, as well as the commercial and monetizable value created by the trading system (Young 2015).

The Role of Pricing in Cities and on Farms

In cities, water pricing tends to be the simplest and most effective tool for compensating the service provider. At the same time, high water prices generally work in reducing city demand, and targeted subsidies or bloc tariffs can be strategically employed to ensure that the most vulnerable residents retain access to affordable water. When utilities need to recover costs through pricing, they also have an incentive to prevent wastage and revenue losses by fixing leaks in the system (Olmstead, Hanemann, and Stavins 2007). In fact, a staggering 32 billion cubic meters of treated water is lost from urban systems around the world each year through leaks in the pipes. Half of these losses occur in developing countries, where customers frequently suffer from supply interruptions and poor water quality (Kingdom, Liemberger, and Marin 2006). Further, when water is priced appropriately, water utilities become beholden to their customers for generating revenue, rather than to political interests for providing subsidies. This increases their incentives to expand service and quality throughout cities, rather than to only politically connected communities.

Utility cost recovery is also important for ensuring that utilities can secure adequate financing. Private financiers are reluctant to invest in utilities that are not self-standing and rely on government subsidies to stay afloat. As a result, private financing is unavailable to many utilities, or it must be backed by public guarantees, greatly reducing the utilities' ability to invest in upgrading or expanding their infrastructure. In the developing world, cost recovery rates are abysmally low; in 2004, 89 percent of utilities in low-income countries and 37 percent of utilities in lower-middle-income countries charged tariffs that were too low to cover basic operation and maintenance costs (Komives et al. 2005), and little has changed since then. Closing this gap could greatly improve the ability of utilities to make investments that increase access to and the reliability of piped water.

Municipal water use is only a small fraction of total water use; significantly more water flows to irrigated farmland. Pricing agricultural water is more complex because its demand tends to be more inelastic than municipal water—farmers need a set volume and tend to pay the asking price, up to a limit. This makes pricing less effective in reducing demand (Molle and Berkoff 2007). Further, agricultural water use is much more difficult to track than municipal water, particularly when farmers can drill into an aquifer to access groundwater, thus circumventing higher prices charged for surface water (World Bank 2010).

Water Trading Schemes—An Option Whose Time Has Come for Consideration, If Not Immediate Implementation

An alternative solution is the use of water permits or water trading schemes. These are systems that give users the right to "sell" or "rent" the water that is available to them. The result is a win-win because a transfer will occur only if the buyer and seller both anticipate a benefit from the transaction. The challenge to establishing such systems is the complex legal and institutional architecture that they need for credibility. If this can be achieved, then market trades will naturally emerge in ways that deliver water to higher-value uses; and with proper regulation, the system could ensure greater sustainability and access for all. Since countries differ in legal systems, institutional capacity, social norms, and economic status, there is no universal template for establishing effective water trades. For instance, it is unlikely that the legal and institutional architecture that has worked for, say, Australia, with its common law traditions and strong institutions, would work in a developing country with a legacy of conflict, such as Angola. But experience in related contexts (such as fisheries) suggests that there are four key issues for creating a workable trading system; these are summarized in box 5.2.

A water trading system is a powerful economic tool whose time has come for consideration, if not immediate implementation. The litany of issues that need to be addressed to build a system at scale may seem overwhelming, but the costs and consequences of inaction are perhaps even more daunting.

Essentials for a Water Trading System

Four main issues must be addressed when developing a successful water trading system:

Establishing and decoupling water rights. Often, the right to water is tied to the ownership of land. It is self-evident that to promote water-related trade and transfers, the right to land ownership must be decoupled from the right to use and trade water. This may be complicated by cultural values or legal constraints and may require significant resources and coordination between jurisdictions. The costs of formally defining water rights can also be substantial, so the scope of the reform needs to be consistent with the availability of resources and institutional capacity.

Standards and measurement. Trade will go smoothly if there are formal and acceptable standards of measurement and accounting for both withdrawals and return flows (which is the "unused" water that is returned to the basin), together with a system of monitoring and estimating these flows. Providing a verifiable measure of the quantities of water that are traded is a prerequisite for lowering the transaction costs of trade. Such standards enable the development of risk mitigation instruments, such as contingent future contracts, and provide credible evidence when legal disputes arise.

Externalities. Water transfers across large geographical areas, or in large quantities, will alter the flows that are returned to the basin. This calls for establishing an environmentally determined limit on the amount of water that is consumed—that is, withdrawn and not returned to the source—as well as standards regarding the quality of return flows. The science required to determine environmental flow regimes has evolved considerably, but debates still persist on methodologies and the practicality of applying different approaches.

Trade volumes and infrastructure. Markets work best when they are competitive, which implies the need for systems where numerous sellers and buyers can interact. However, this can only occur if there is sufficient infrastructure that can connect sellers to buyers. The cost to each market participant for building, maintaining, and financing such infrastructure may render a trading system unaffordable. But since much of this infrastructure confers public good benefits, and there are likely to be coordination constraints, there is a prima facie case for government subsidies to meet a fraction of the capital costs.

Informal Water Markets Are Pervasive but Have Limitations

Since water is so essential for all activities, when official systems fail to deliver water, informal markets have emerged to fill the gap. This is most visible in cities, where informal vendors haul water along the streets, but is also prevalent in rural areas. Rural India and Pakistan, for instance, have flourishing local markets, where farmers engage in bilateral trading to buy and sell surface water and groundwater. In Oman, the *aflaj* system facilitates trade between villages. Even in Spain, informal transactions are common at the village level (Palomo-Hierro, Gómez-Limón, and Riesgo 2015).

Lacking legal status, these transactions must rely on social norms, trust, and power relations for contract enforcement and are therefore limited in scope and scale. Further, when these systems break down, and the supply of water becomes monopolized, pricing can become extortionary. The outcome is that the urban poor, who are unserved or underserved by water utilities, often pay a much higher price to water vendors than the rich, who have access to piped water.

Improving Efficiency—Useful, but No Panacea

Improving the efficiency of water use is always popular among policy makers because it offers the tantalizing prospect of generating "new" water through efficiency savings, without depriving any existing users. Efficiency gains can be obtained by producing more output per unit of water consumed, either with prevailing techniques or through the adoption of new technologies. Since water is typically underpriced, there is a presumption that the savings generated through improvements in efficiency could be significant—either by bringing firms closer to the efficiency frontier or by encouraging the adoption of new technologies.

There is a wide range of new technologies and techniques for water saving, including practices like alternate wetting and drying (AWD) for rice crops, or intensification of dryland agriculture. Studies show that advanced irrigation technologies, such as subsurface drip irrigation and micro-irrigation, can substantially improve crop yields while reducing total water consumption (Ayars, Fulton, and Taylor 2015). However, adoption of these solutions is slow, hesitant, and below desired levels. The constraints most often lie in misaligned incentives. A large proportion of the benefits of efficiency improvements are public, while technology adoption costs are private. This requires sharper incentives for technology uptake, which might necessitate a change in the subsidy regime, public investments in infrastructure, or increasing access to credit. The appropriate policy response depends on the context and needs to be designed to avoid perverse incentives that discourage investments, either because of moral hazard or heightened transaction costs (Malik, Giordano, and Rathore 2016).

Not everyone agrees that such policies would effectively manage demand, and there is much debate in the water resources literature on the use of efficiency measures, especially for irrigation. Critics contend that improvements in water efficiency have often failed to deliver the anticipated water savings. This is likely to be true in two circumstances.

First, there is the *Jevons Effect*, a well-known result in resource economics, named after the 19th-century economist Stanley Jevons. The Jevons Effect suggests that improvements in the efficiency of resource use may not necessarily translate into greater savings of the resource and could even result in more of it being used. This seemingly perverse outcome occurs when the water that is "saved" through efficiency improvements, is used to expand production[5]—a phenomenon termed either the "rebound" effect or, more precisely, an "output" effect. It is critical for policy makers to determine whether the output effect is always large enough to offset the savings from improved efficiency. This is most likely to occur when demand for the final product is sufficiently price elastic,[6] or when the resource (water) is a binding physical constraint on production.[7]

Another way in which water savings could be eroded is when a new technology alters the magnitude of return flows. For instance, drip irrigation may lead to less water being withdrawn from a river. But if the new technology means that crops consume more water, then less could be returned to the system, harming downstream users. Thus, the key policy question is whether the decline in return flows wipes away benefits of lower amounts of water being withdrawn. These are ultimately empirical questions that would need to be assessed for the problem under consideration.

Complementary Policies to Protect the Vulnerable from Water Shocks

Water is a merit good that is necessary for life and is therefore classified as a "human right" by the United Nations. When water scarcity combines with rainfall variability, poor and vulnerable populations often go unsupplied or undersupplied, leading to long-lasting and often unseen consequences. The World Meteorological Organization cautions that it is easy to miss the onset of a drought when the baseline variability of rainfall is high.[8] It is also easy for drought impacts to be amplified through maladaptations that increase vulnerability and the overuse of scarce water. The hidden effects of a drought that linger long after the drought itself has retreated are of greatest concern, as a result of their irreversible and intergenerational impacts. Addressing these risks calls for complementary policies and interventions.

Safety net programs, such as cash and in-kind transfers, have long been used to provide a minimum level of support to those in need, both during normal times and in times of crisis. Those programs are also increasingly combined with measures aimed at improving human development outcomes, for instance, in nutrition and child care. Such programs can protect livelihoods, reduce transitory and chronic poverty, and eliminate food insecurity. Addressing the types of shocks and irreversible impacts identified in this book would mean implementing programs that target vulnerable populations most in need. In practical terms, this might mean designing programs that cover only the poorest of the population, or those that are chronically food-insecure, with geographic targeting in areas where rainfall shocks are most frequent. These approaches will always be challenging because of risks of capture, exclusion (counting as nonpoor individuals or households that are), leakage (counting as poor those that are not), limited knowledge, and capacity constraints. Safety net benefits can also be universal, available to all households, eliminating some of the difficulty of targeting households. Not surprisingly, universal subsidies are costly, and the fiscal strains associated with them erode the level of program benefits and their effectiveness. Hence, most programs in developing countries are targeted. But a recent review of safety nets in developing countries found that targeting is typically imprecise and few programs are set up to assist households managing idiosyncratic shocks (Monchuk 2013). There is, however, a growing recognition of the potential for safety nets to increase households' resilience to climate variability, for instance, in the Sahel region with the Sahel Adaptive Social Protection Program.[9]

Development and inclusive growth are perhaps the best antidotes to a natural disaster. In the meantime, an effective social safety net would go a long way toward protecting people against the rainfall shocks that will inevitably intensify with climate change. But when these are deemed to be unaffordable or unfeasible, an insurance-based approach may be useful. Determining insurance payout triggers and amounts is often difficult, especially in poor rural areas where governments or private insurance companies may not have the capacity to survey. One solution is a weather index insurance scheme, where payouts are based on triggers that are correlates of losses, rather than actual losses of individuals or area losses themselves. For example, payment of a known sum might be triggered by satellite data on rainfall or satellite imagery measuring the blueness of the ground as a proxy for the degree of flooding.

An advantage of this approach is that in times of need, there is an automatic payout with little room for fraudulent manipulation. The main disadvantage is the risk that a payout might not be made even though a loss has occurred, or that a payout might be made even when there is no loss. Given the depth of the consequences of delaying action to mitigate the impact of water shocks, the risks of errors of exclusion (triggering insurance in the absence of a disaster) may well be outweighed by the benefits of a system that is wide enough to include those experiencing water shocks. More research could provide better information on these relative costs and risks.

The results that link rainfall shocks and deforestation, which were discussed in chapter 2, heighten the need to put suitable safety nets in place. This dynamic demonstrates that rainfall shocks impose costs not just on the farmer, who sees reduced yields, but also on the local region that loses valuable forested land, and on the world that sees a rise in carbon emissions. If farmers were compensated by a safety net program or insurance scheme, they would have less of an incentive to expand their cropland into forested land, which does require a large fixed cost of both time and money. This implies that an economically efficient program should compensate farmers more if they farm land adjacent to forests, perhaps with the conditions that they do not encroach on the forested land and that they lived there prior to the program becoming active (to prevent people from moving adjacent to forests simply to receive the added benefits).

While there is no perfect scheme to buffer poor and vulnerable populations from rainfall shocks, there is clearly a need for complementary policies that provide a buffer in times of crisis. The choice of approach will depend on country-specific circumstances and on climate and weather risks, as well as on fiscal and administrative capacity. In circumstances where there is chronic poverty and food insecurity, priority should go to approaches that target highly vulnerable households. In other circumstances where poverty is less acute, identifying short-term shocks as the trigger for payment may be more appropriate.

Policies to Protect the Sources of Water

Since water is a renewable natural resource, the degradation of aquatic ecosystems has the potential to diminish the current and future productivity of water. Appropriate stewardship and management of water resources will become more important as the demand for water continues to grow and pressures on aquatic ecosystems multiply. Adequate safeguards, such as quotas and water quality standards, are required to protect water sources, and to prevent overuse and abuse of these public goods.

The volume of water available to humans scales approximately with the area of watersheds left on the planet. Watersheds collect, store, and filter water, and, when managed well, provide a number of additional benefits to people and nature. The supply of surface water resources is therefore only renewable if watersheds are maintained and conserved in ways that allow them to continue to collect water from the atmosphere and steadily release it into rivers and on land from where it can be used. This essential point is vastly underappreciated in policy discussions about water and its sustainable use.

The quality (chemical and thermal properties) of water is as important as its quantity. Rivers have been used for millennia as a way of disposing of sewage and waste. With economic growth, household waste has been complemented by agricultural and industrial discharges. Reducing nutrient loads from municipal sources with known technologies has proven to be comparatively simple. Developing effective strategies to deal with agricultural nonpoint sources and persistent hazardous waste is much more difficult. Effective management of these requires coordinating and controlling behavior across diverse economic actors and ecological landscapes.

Balancing ecosystem sustainability with immediate economic pressures remains among the most challenging tasks for policy makers, particularly when it comes to determining how much water is made available for the environment and determining what water infrastructure will be built and how it will be operated. The appropriate policy mix will need to vary with ecological pressures and country circumstances. Each waterway is distinctive in flow regime, and countries vary in their capacity to implement policies. Available policies range from quantity controls such as limits on withdrawals, to pollution regulations, to more sophisticated economic instruments that seek to alter incentives and behavior.

Conclusions

There is no single solution to the myriad problems of managing the impacts of water. Still, better incentives, improved ownership rights, and smart supply options can open the door to solutions. Many schemes around the world have attempted one or more of these policies in a piecemeal manner; what is required is a better-coordinated effort.

Success will depend on the ability to address institutional and governance barriers that can be politically and economically perplexing. Successful efforts to improve water management may involve a "ladder of interventions"—a set of policy and institutional responses that become increasingly complex as pressures and capacity grow. It is critical for local and national governments, multilateral institutions, civil society, and water users from all sectors and backgrounds to engage in sustained dialogue to identify priorities for water allocation and the processes that will build a consensus.

In the end, water allocation is the key to maximizing water's value for poverty reduction, growth, and sustainable development. Though the task may seem difficult, the costs of neglect and inaction greatly outweigh the challenges associated with establishing prudent stewardship and management. With the right policies and determined priorities, the risks of fickle rainfall can be converted into an opportunity for purposeful progress.

Notes

1. The ancient Greeks believed that Zeus unleashed a flood because of his anger with humans. A more complicated flood legend appears in the even more ancient *Avestan* texts of Persia. In Hindu mythology, a flood is associated with Brahma in the *Tale of Manu*, and there is the flood in the biblical narrative of Noah's Ark that also appears in *Safina Nuh*, the Islamic rendering of the ark (https://en.wikipedia.org/wiki/Noah).

2. http://www.gogeomatics.ca/magazine/mcgill-geography-researchers-map-global-impact-of-dams.htm#

3. http://css.umich.edu/sites/default/files/U.S._Water_Supply_and_Distribution_Factsheet_CSS05-17.pdf.

4. The correlation between area irrigated and water scarcity has long been recognized. For instance, Scheierling and Treguer (2016) provide examples such as Saudi Arabia, where the area irrigated increased from 0.3 to 1.6 million hectares (ha); Libya, from 0.1 to 0.5 million ha; India, from 26 to 67 million ha; and the Republic of Yemen, from 0.2 to 0.7 million ha. The increases in each case correlated with higher levels of water scarcity. There are of course other factors at play, such as rising populations and GDP, but agriculture remains the main user of water.

5. This is highly desirable in many contexts.

6. To see why, consider a partial equilibrium situation where water is used as an input for a product. If product demand is highly elastic, and the efficiency gains lower production costs and therefore prices, then product demand will expand. Since demand is elastic, it will expand more than the fall in price, potentially inducing the Jevons Effect.

7. Suppose that water use is costless and the water availability constraint binds; then relaxing the resource constraint will in most cases lead to an expansion in output and water use.

8. http://cla.auburn.edu/ces/climate/droughts-and-floods/.

9. http://www.worldbank.org/en/programs/sahel-adaptive-social-protection-program-trust-fund.

References

Abell, R., N. Asquith, G. Boccaletti, L. Bremer, E. Chapin, A. Erickson-Quiroz,. J. Higgins, J. Johnson, S. Kang, N. Karres, B. Lehner, R. McDonald, J. Raepple, D. Shemie, E. Simmons, A. Sridhar, K. Vigerstøl, A. Vogl, and Sylvia Wood. 2017. *Beyond the Source: The Environmental, Economic and Community Benefits of Source Water Protection.* Arlington, VA: Nature Conservancy.

Annandale, George W., Gregory L. Morris, and Pravin Karki. 2016. "Extending the Life of Reservoirs: Sustainable Sediment Management for Dams and Run-of-River Hydropower." Washington, DC: World Bank.

Ayars, J. E., A. Fulton, and B. Taylor. 2015. "Subsurface Drip Irrigation in California—Here to Stay?" *Agricultural Water Management* 157: 39–47.

Buchauer K., and C. Huang. 2016. "Recommendations for Urban Sludge Management in China." Working Paper, World Bank, Washington, DC.

Chen, S., and B. Chen. 2013. "Net Energy Production and Emissions Mitigation of Domestic Wastewater Treatment System: A Comparison of Different Biogas–Sludge Use Alternatives." *Bioresource Technology* 144: 296–303.

Damania, R., A. Glauber, P. Scandizzo, T. Von Platen-Hallermund, A. Barra, D. Aryal, and M. Haji. 2015. "Tanzania's Tourism Futures: Harnessing Natural Assets." Environment and Natural Resources Global Practice Policy Note, World Bank, Washington, DC. http://documents.worldbank.org/curated/en/204341467992501917/Tanzania-s-tourism-futures-harnessing-natural-assets.

Grafton, R., D. Garrick, and J. Horne. 2017. *Water Misallocation: Governance Challenges and Responses.* Draft Policy Report, World Bank, Washington, DC.

Guannel, G., K. Arkema, P. Ruggiero, and G. Vertues. 2016. "The Power of Three: Coral Reefs, Seagrasses and Mangroves Protect Coastal Regions and Increase Their Resilience." *PLoS One*, July 13.

Kingdom, B., R. Liemberger, and P. Marin. 2006. "The Challenge of Reducing Non-Revenue Water (NRW) in Developing Countries: How the Private Sector Can Help—A Look at Performance-Based Service Contracting." Water Supply and Sanitation Board Discussion Paper Series, Paper 8, World Bank, Washington, DC.

Komives, K., V. Foster, J. Halpern, and Q. Wodon. 2005. "Water, Electricity, and the Poor: Who Benefits from Utility Subsidies?" World Bank, Washington, DC.

Malik, R. P. S., M. Giordano, and M. S. Rathore. 2016. "The Negative Impact of Subsidies on the Adoption of Drip Irrigation in India: Evidence from Madhya Pradesh." *International Journal of Water Resources Development.* October 13.

Mekonnen, M., and A. Hoekstra. 2016. "Four Billion People Facing Severe Water Scarcity." *Science Advances* 2 (2): e1500323.

Menzen-Dick, R., and C. Ringler. 2008. "Water Reallocation: Drivers, Challenges, Threats, and Solutions for the Poor." *Journal of Human Development* 9 (1): 47–64.

Molle, F., and J. Berkoff, eds. 2007. *Irrigation Water Pricing: The Gap between Theory and Practice.* Vol. 4. CABI.

Monchuk, V. 2013. *Reducing Poverty and Investing in People: The New Role of Safety Nets in Africa.* Directions in Development. Washington, DC: World Bank.

Olmstead, S., W. Hanemann, and R. Stavins. 2007. "Water Demand under Alternative Price Structures." *Journal of Environmental Economics and Management* 54 (2): 181–98.

Palomo-Hierro, S., J. Gómez-Limón, and L. Riesgo. 2015. "Water Markets in Spain: Performance and Challenges." *Water* 7 (2): 652–78.

Potter, G. 2014. "Agricultural Subsidies Remain a Staple in the Industrial World." Vital Signs, February 28. Worldwatch Institute, Washington, DC. http://vitalsigns. worldwatch.org/sites/default/files/vital_signs_trend_agricultural_subsidies_final _pdf_0.pdf.

Scheierling, S. M., and D. O. Treguer. 2016. "Enhancing Water Productivity in Irrigated Agriculture in the Face of Water Scarcity." *Choices* 31 (3).

Watson, K. B., T. Ricketts, G. Galford, S. Polasky, and J. O'Niel-Dunne. 2016. "Quantifying Flood Mitigation Services: The Economic Value of Otter Creek Wetlands and Floodplains to Middlebury, VT." *Ecological Economics* 130: 16–24.

World Bank. 2010. *World Development Report 2010: Development and Climate Change.* Washington, DC: World Bank.

Young, M. 2015. "Doubling the Value of Water in the American West." *Water Economics and Policy* 1 (4): 1571003.

Zarfl, C., A. E. Lumsdon, J. Berlekamp, L. Tydecks, and K. Tockner. 2015. "A Global Boom in Hydropower Dam Construction." *Aquatic Sciences* 77 (1): 161–70.

Appendix A

This brief appendix describes the rainfall data used in this book, and compares the measures used in this book with other well-known rainfall indices.

Weather Data

The raw weather data used in this book come from Willmott and Matsuura (2001) from the University of Delaware (henceforth referred to as the "Delaware data").[1] The spatial coverage of the Delaware data is the 0.5 degree gridcell level (approximately 50 square kilometers at the equator), centered on the 0.25 degree, for all major land masses (excluding Antarctica). Temporally, it is available monthly from 1901 until 2014. The gridded data set is generated from weather station data compiled from several different sources. These station data are then converted into gridded data using a combination of several different spatial interpolation methods, including digital-elevation-model assisted interpolation, traditional interpolation, and climatologically aided interpolation.[2]

The Delaware data have several advantages. The global spatial coverage is vital for a global study such as this, and avoids the need to use different data sources for different countries or regions, which may not be comparable. The lengthy temporal dimension is also highly desirable, as this study uses data covering eight decades, and looks back even further in time to calculate long-run means. The data are also regarded as being highly reliable and are used quite often in the economics literature.[3] Last, the Delaware data set provides measures of both rainfall and temperature, which are calculated using a similar methodology. Because temperature and rainfall are highly correlated, it is important that temperature is controlled for in all of the regressions that use rainfall. And as spatial interpolation can create systematic biases, it is important that both weather measures are calculated using similar methodologies.

Defining Rainfall Shocks

The rainfall shocks defined in this book are based on local, long-term conditions. In order to define when a region experienced a shock, the *z-score* of rainfall is calculated in each gridcell, and each time period:

$$z_{it} = \frac{precip_{it} - \overline{precip_i}}{\sigma_i};$$

where z_{it} is the z-score of rainfall in gridcell i at time t, $precip_{it}$ is rainfall in gridcell i at time t, $\overline{precip_i}$ is mean rainfall in gridcell i over the full time period of the Delaware dataset (1901–2014), and σ_i is the standard deviation of rainfall in gridcell i over the same time period. The z-score for rainfall is interpreted as the number of standard deviations away from the long-run mean in time t. The length of time t takes several different values throughout the book depending on data availability and what is being tested. In some specifications, t is a year, and in other specifications, t is a month. In the case of monthly z-scores, the long run mean rainfall, and its standard deviation is calculated based on historical means of the same month. For instance, in gridcell i, each month will have its own independent long-run mean and standard deviation based on historical values for that particular month.

Once z-scores are calculated, binary variables are generated to indicate whether the z-score passes certain thresholds. Depending on what is being tested, these thresholds are z-scores of either ±1 (considered to be moderate shocks), or ±2 (considered to be large shocks). Doing this allows for a semiparametric testing of the impacts of deviations from the long-run mean. It allows for impacts to be both nonlinear, and also nonsymmetric around zero.[4] These properties are important for distinguishing the possible heterogeneous impacts of deluges versus droughts.

For the majority of the research in the book, the empirical strategy is similar to that of a natural experiment. Outcomes in regions during time periods when there were no rainfall shocks are compared to times when there was one. Fixed effects and control variables are carefully chosen to control for other trending or cyclical factors that may influence the relationship between rainfall shocks and the studied outcomes. Importantly, identification is predicated on the fact that rainfall shocks are exogenous and unpredictable. In order to satisfy these conditions, rainfall shocks must be stationary over time. To test for this, a Levin-Lin-Chu panel unit root test was performed on several of the rainfall and shock indicators. Results presented in table A.1 conclusively show that both the raw rainfall data and the shock variables are highly stationary.

Testing Annual Rainfall and Rainfall Shocks for Stationarity

	Levin-Lin-Chu Unit Root Test Statistic (p Value)	
Precipitation indicator	1901–2014	1980–2014
Precipitation (mm/year)	−1,100 (0.0000)	−500 (0.0000)
Annual +1 SD Shock	−1,300 (0.0000)	−530 (0.0000)
Annual +2 SD Shock	−1,400 (0.0000)	−150 (0.0000)
Annual −1 SD Shock	−1,200 (0.0000)	−460 (0.0000)
Annual −2 SD Shock	−780 (0.0000)	−520 (0.0000)

Note: The table shows *t* statistics (*p* values) from the Levin-Lin-Chu (LLC) unit root test performed on the raw, annual rainfall data from the Delaware data, and the shock indicators that were developed from this data, as described above. Column 1 shows *t* statistics from the full time period, 1901–2014, while column 2 shows t-statistics from a shortened, more recent time period, 1980–2014. The Augmented Dickey-Fuller regression that is fitted for each panel includes 1 lag, chosen based on the Akaike Information Criterion. The null hypothesis of the LLC test is that the panels contain unit roots. Rejection of the null hypothesis implies that the panels are stationary. The tests confirm that the rainfall data, and all of its derivatives, are highly stationary, giving further confidence in the validity of the results that use this data.

Rainfall Shocks' Relatability to Other Extreme Events and Indices

Correctly identifying extreme events is very often not a straightforward exercise. Floods, and their consequences, are observable, making their identification somewhat easier. The intensity of floods can be characterized by the level of water, the number of fatalities they generate, the number of people they displace, and the economic cost of destruction and disruption they cause. Categorizing droughts, however, is much more complicated. Their damages are less visible. One may observe reservoir levels declining, but this process is much more gradual.

Even more challenging than identifying these extremes is identifying moderate variations in water availability, which may be just as useful for farmers, public utility managers, policy makers, or researchers. These moderate "shocks" are less visible, and their impacts may be highly heterogeneous depending on a whole host of factors such as geography, demographics, and climate. Meteorologists have developed several indexes to identify and characterize abnormal rainfall events, whether they are dry or wet, moderate or extreme. Three of the more widely used indices are described here.

The Palmer Drought Severity Index

The Palmer Drought Severity Index (PDSI), developed by Palmer in 1965, is among the earliest and the most influential indexes on droughts. It aims to quantify the gap between the available level of water in a given location and the required level of water necessary to support a near-normal level of economic activity (Dai and National Center for Atmospheric Research Staff 2015; Palmer 1965). To that end, the PDSI takes into account precipitation, evapotranspiration, and soil moisture conditions, to model a more accurate total water balance. This index can be very useful when calibrated for a precise area (Palmer's work originally focused on the Midwest region of the United States). However, the approach has proved more difficult to use at larger scales. Indeed, data on soil moisture, a necessary component of the index, is scarce. In addition, it has been shown that the PDSI is highly sensitive to the assumptions made to calibrate the model; many of these assumptions are still debated by meteorologists (Alley 1984).

The Standardized Precipitation Index

More recently, an alternative index—the Standardized Precipitation Index (SPI)—has become more widely used, partly to overcome some of the problems inherent to the PDSI. The SPI takes into account rainfall levels only. Based on the local history of rainfall, it determines the probability of observing a rainfall event of a given magnitude (Kayantash and National Center for Atmospheric Research Staff 2015). Historical rainfall at each location is first fitted to a probability distribution function (generally, a gamma distribution), that is then normalized. From the normalized index, the standard deviation of rainfall with respect to the long-term mean of rainfall is calculated. The SPI is flexible and can be calculated for different time scales (often between 1 and 36 months).

The rainfall shocks calculated in this book are most similar to the SPI. To test the robustness of the results to other rainfall shock measures, a panel of SPI was calculated from the Delaware data for this book. The z-scores described above are found to be highly correlated with the SPI (>0.98 correlation coefficient), and regressions estimated with shocks generated from the SPI are very similar to the results presented in the book. These results can be found in the background papers in Volume 2 of the book.

The Standardized Precipitation Evapotranspiration Index

Because the SPI and the PDSI are based on two very different methodologies, they differ in what they qualify as droughts or wet spells (Guttman 1998). A more recent index, the Standardized Precipitation Evapotranspiration Index (SPEI), constitutes a middle ground between the two indexes by reintegrating evapotranspiration into the SPI. As with the PDSI, however, this index is very sensitive to the chosen definition of evapotranspiration (Vicente-Serrano and National Center for Atmospheric Research Staff 2015). Selected results from this book have been tested against the SPEI and the main conclusions were found to be robust to the use of this alternative measure.[5]

Notes

1. Data provided by the NOAA/OAR/ESRL PSD, Boulder, Colorado, from their website at http://www.esrl.noaa.gov/psd/, in any documents or publications using these data.

2. For information on weather station data sources and interpolation techniques, see http://climate.geog.udel.edu/~climate/html_pages/Global2014/README. GlobalTsP2014.html for rainfall, and http://climate.geog.udel.edu/~climate/ html_pages/Global2014/README.GlobalTsT2014.html for temperature.

3. For a discussion of the Delaware dataset and a comparison with other frequently used weather datasets in the economics literature, see Auffhammer et al. 2013.

4. The cutoffs of ±1 and ±2 are somewhat arbitrary. A sensitivity analysis was performed around these values and found the results to be consistent. Results are available from the authors upon request.

5. See chapter 2 of this book.

References

Alley, W. 1984. "The Palmer Drought Severity Index: Limitations and Assumptions." *Journal of Climate and Applied Meteorology* 23 (7): 1100–09.

Auffhammer, M., S. M. Hsiang, W. Schlenker, and A. Sobel. 2013. "Using Weather Data and Climate Model Output in Economic Analyses of Climate Change. *Review of Environmental Economics and Policy* 7 (2): 181–98.

Dai, A., and National Center for Atmospheric Research Staff, eds. 2015. "The Climate Data Guide: Palmer Drought Severity Index (PDSI)." National Center for Atmospheric Research, Boulder CO. https://climatedataguide.ucar.edu/climate-data /palmer-drought-severity-index-pdsi.

Guttman, N. 1998. "Comparing the Palmer Drought Index and the Standardized Precipitation Index." *Journal of the American Water Resources Association* 34 (1): 113–21.

Kayantash, J., and National Center for Atmospheric Research Staff, eds. 2015. "The Climate Data Guide: Standardized Precipitation Index (SPI)." National Center for Atmospheric Research, Boulder CO. https://climatedataguide.ucar.edu/climate-data /standardized-precipitation-index-spi.

Palmer, W. 1965. *Meteorological Drought*. Vol. 30. Washington, DC: US Department of Commerce, Weather Bureau.

Vicente-Serrano, S., and National Center for Atmospheric Research Staff, eds. 2015. "The Climate Data Guide: Standardized Precipitation Evapotranspiration Index (SPEI)." National Center for Atmospheric Research, Boulder, CO. https://climatedataguide. ucar.edu/climate-data/standardized-precipitation-evapotranspiration-index-spei.

Willmott, C. J., and K. Matsuura. 2001. "Terrestrial Air Temperature and Precipitation: Monthly and Annual Time Series (1900–2014)." http://climate.geog.udel.edu/~climate /html_pages/download.html.